PENGUIN BUSINESS
PURPOSE

Pankaj Setia conducts executive education programmes and consults with leading organizations and government institutions. His opinions on digital transformation and leadership have appeared in key publications, such as *Businessworld*, *Outlook* and *Forbes India*, among others. Pankaj teaches graduate-level courses on the leadership of digital organizations, strategic management of digital innovations and digital transformation at the Indian Institute of Management Ahmedabad. He has taught for many years at Michigan State University and the University of Arkansas in the US. He also offers free courses on digital transformation on Coursera.

His research has been published in top information systems and business journals, such as *MIS Quarterly*, *Information Systems Research* and *Journal of Operations Management*, among others. He is often quoted by national dailies, such as *Economic Times*, *Indian Express* and many others, and has featured among the top 100 researchers in the field of information systems worldwide.

ADVANCE PRAISE FOR THE BOOK

'This fascinating and well-researched book unravels the deep linkages between organizations, life and technologies. A must-read for those interested in driving successful digital transformations'—**Saurabh Garg, Secretary, Ministry of Statistics and Programme Implementation, and former CEO, UIDAI**

'Pankaj Setia's book is a must-read for anyone who is seeking to evangelize, lead, organize or manage the digital transformation of any organization. The book offers both a strongly grounded conceptual framework as well as a pragmatic guide to think about digital transformation from multiple points of view. This will be particularly relevant as organizations seek to navigate AI-led transformation'—**Prof. Ramayya Krishnan, dean, Heinz College of Information Systems and Public Policy, and William W. and Ruth F. Cooper Professor of Management Science and Information Systems, Carnegie Mellon University**

'This book is a must-read for anyone seeking to navigate the complex terrain of digital transformation. Prof. Pankaj Setia's ability to connect the dots between neuroscience, technology and human behaviour is truly remarkable. This book goes beyond the superficial aspects of digital transformation and delves into its latent purpose, offering a fresh perspective on how organizations and leaders should approach the digital age. The book is not just about adopting new tools but also about understanding how those tools reshape human experience, decision-making and organizational structure. It brings forward a percipient insight into the future of work, society and technology—one that emphasizes both technical proficiency and emotional intelligence equally. For anyone researching or leading digital transformation, this book provides essential foundational principles that, when integrated with the digital ecosystem, will shape strategies and thinking for years to come'—**Kaku Nakhate, president and country head (India), Bank of America**

'The strategic shifts in telecommunication, computing, artificial intelligence and social media are continuing to change the way we work, play and live. Over the ages, technology has generally benefited mankind. However, its increasing overlaps with human cognitive abilities have led to some apprehensions. In his book *Purpose*, Prof. Setia presents a well-researched, simply written text to clarify what digital transformation will mean for society. He rightly suggests that early adoption of digital transformation will lead to substantial benefits for mankind. I strongly recommend this book for people seeking a clear understanding of digital transformation and its effects on their lives and society'—**Aditya Puri, senior adviser, The Carlyle Group, and former CEO and MD, HDFC Bank**

'*Purpose* is a timely and insightful exploration of how digital transformation is reshaping individuals, organizations and societies. Pankaj Setia provides a deeply thoughtful and scientifically grounded perspective on the forces driving this transformation. His unique blend of expertise, drawn from decades of research, offers readers a framework to not only understand but also harness digital technologies in a purposeful way. This book is a must-read for anyone looking to navigate the complexities of the digital age and unlock its full potential for growth and innovation'—**Abhishek Singh, additional secretary, MeitY, and former CEO, Digital India Corporation**

PURPOSE

Digital Transformation
of Individuals, Organizations
and Societies

PANKAJ SETIA

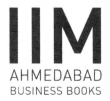

AHMEDABAD
BUSINESS BOOKS

PENGUIN
BUSINESS

An imprint of Penguin Random House

PENGUIN BUSINESS

Penguin Business is an imprint of the Penguin Random House group of companies whose addresses can be found at global.penguinrandomhouse.com

Published by Penguin Random House India Pvt. Ltd
4th Floor, Capital Tower 1, MG Road,
Gurugram 122 002, Haryana, India

First published in Penguin Business by Penguin Random House India 2024

Copyright © Pankaj Setia 2024

All rights reserved

10 9 8 7 6 5 4 3 2 1

The views and opinions expressed in this book are the author's own and the facts are as reported by him which have been verified to the extent possible, and the publishers are not in any way liable for the same.

Please note that no part of this book may be used or reproduced in any manner for the purpose of training artificial intelligence technologies or systems.

ISBN 9780143470076

Typeset in Adobe Caslon Pro by Manipal Technologies Limited, Manipal

This book is sold subject to the condition that it shall not, by way of trade or otherwise, be lent, resold, hired out, or otherwise circulated without the publisher's prior consent in any form of binding or cover other than that in which it is published and without a similar condition including this condition being imposed on the subsequent purchaser.

www.penguin.co.in

Contents

Preface ix

1. Digital Organization: The Digital
   Transformation of an Organization 1

Part 1: Digital Transformation

2. Digital Transformation of an Organization:
   The Scientific Foundations 21

Part 2: The Instrumental Purpose of Digital Transformation

3. The Instrumental Purpose of an Organization's
   Digital Transformation 37

4. Visualizing and Conceptualizing Digital Capabilities — 51

Part 3: The Operational Purpose

5. Operationalizing Digital Transformation — 71
6. Transforming Digital Architecture — 81
7. Transforming Work — 107
8. Transforming Governance — 131
9. Transforming Business Model — 151
10. The Extended DaWoGoMo© Model — 165

Part 4: The Existential Purpose

11. The Existential Purpose: Life and Digitization — 175
12. Revisiting the Instrumental Purpose: Digital Transformation and Individuals — 190
13. Operational Purpose Revisited: The Force Underlying DaWoGoMo© + Culture — 210
14. The Existential Purpose: Computational Freedom — 228
15. Digital Organization — 244

Part 5: Enacting a Digital Transformation

16. The Purpose — 267

Notes — 283

Preface

This book reveals the latent purpose required to champion, lead and enact digital transformations. I articulate that the purposeful approach followed in this book is urgently required and may help the world's population (approximately 800 crore people) harness the power of digital technology. This book also articulates the rapidly emerging narrative that digital technologies are currently governing the world or will soon do so. Ironically, despite these developments, the purpose of digital transformations is not well articulated by most, if at all. So, what makes digital technologies and the organizations championing them dominant? Two factors explain their dominance: they can do a) what humans ordinarily cannot or b) what humans can do but not that effectively. While discussing digital technologies, I do not refer to artificial intelligence

(AI) alone. Technologies that existed before the advent of AI performed the same two functions. However, only a few people can understand or talk about transformation. Few understand the *purpose* behind the technology's potent force. I argue that purpose is a latent force, similar to the forces of gravity, electromagnetic forces and other natural forces. The purpose prevails over mundane thinking and weak conceptualizations, making technology central to human existence.

The book is based on over twenty years of my research, thinking, discussions, publications and observations about how technology influences us. I also build upon my understanding of the sciences—physics, chemistry, neuroscience and others—to bring out the notion of purpose—the purpose of digital transformation. The urgency in understanding the purpose of digital transformation stems from the fact that the current trajectory of global development and evolution does not fully align with the rapid advancement of technology. Specifically, the world operates under rigid and strict rules. For example, many of my colleagues pursue literacy goals with a sense of finality. Governments and societies across the world pursue this as a sacrosanct goal. I am not sure this will be so fifty years from now. This statement may make my colleagues in economics and policy cringe. However, as computer chips become embedded in our brains (or will otherwise be linked to our brains), literacy may be just one aspect—and may not be the main one—that requires one to spend many years of one's life. The current global system is based on literacy. However, dynamics may change radically with technologies that can listen ubiquitously or even read minds (literally). Brain–computer interfaces being commercialized by many well-known companies are likely to achieve these feats relatively soon. Large language models may further revolutionize and transform the relevance of writing and literacy in ordinary lives. While reading and writing may remain important, they may not be the most valuable skills for *everyone* to invest years of time in. Alternate ways to live and be productive citizens of the world, society and community may emerge. What would be the purpose of such a society? Not answering the question will likely lead to a lot of problems in the near future.

Many such problems have already started to surface. I identify a transformative paradox to summarize the dilemma. The book is a scientific endeavour that aims to define and address this paradox, highlighting it as a problem stemming from a lack of clarity in purpose. The lack of understanding of purpose may hurt individuals, organizations and societies. To identify *purpose*, I outline complex and interwoven linkages between life, organization and digital technologies. Thinking about digital technologies as an element devoid of links to life is dangerous. We are intertwined with technology today. So, to unravel these interconnections, I espouse a new way to look at ourselves and life using a neuroscientific lens. Further, mastering the principles of organizations is required for transforming digitally. This is because digital transformation is better thought of as *the transformation of life and organization through digital technologies.* In summary, unravelling the linkages between the organization, life and digital technologies, the book outlines a model for purposeful digital transformation. As you read through the book, think deeply about life and organization to unleash the force of purpose.

I appreciate the support from the Institute Chair and the Centre for Digital Transformation at IIM Ahmedabad in writing this book.

1

Digital Organization

The Digital Transformation of an Organization

'Well, that's the universal law of technology, that [it] can be used for good or evil. When humans discovered the bow and arrow, we could use that to bring down game and feed people in our tribe. But of course, the bow and arrow can also be used against our enemies.'

—Michio Kaku, theoretical physicist
and co-founder, string field theory[1]

अद्भिर्गात्राणि शुध्यन्ति मनः सत्येन शुध्यति ।
विद्यातपोभ्यां भूतात्मा बुद्धिर्ज्ञानेन शुध्यति ।।
मनुस्मृति

'Water cleanses the body. The heart is cleansed by the truth. Aatma (soul) is cleansed through education. Intellect is cleansed through knowledge.'

—Manusmriti[2]

This book outlines how to transform digitally. Digital transformation is an endeavour that is changing the meaning of life as it catalyses new (digital) organizations. In our day-to-day lives, digital organizations are engulfing us. Today, in almost all Indian towns and cities, roadside vendors and *kirana* (neighbourhood) stores accept UPI payments (through Paytm, PhonePe, Google Pay and others). You are also digitally transforming by connecting your bank account with these UPI-based apps or payment wallets. Further, many of us are creating innovative digital organizations in our homes with the use of Netflix, YouTube or Jio for entertainment; Alexa or Google Voice for smart living; Coursera, EdX or Online IIMA for learning; Amazon, Flipkart, Alibaba or Ajio for shopping and so on. Creating digital organizations may also entail complex digital transformations involving advanced technologies, such as enterprise resource planning (e.g., SAP), customer relationship management systems (e.g., Salesforce), large language-based models (e.g., ChatGPT) or robotic process automation (RPA) technologies. It is hard to ignore these technological advances because of their potential.

Indeed, the transformative potential of digital technologies is immense. For instance, artificial intelligence (AI)-based technologies are transforming how we write compelling stories and poems, often on par with those written by exceptional journalists and writers. Google created an AI, feeding it over 10,000 books and, in turn, the AI was able to write mournful poems.[3] Robots are now digitizing feelings and beliefs. Telepresence robots add a sense of actual personal presence during meetings.[4] Even the most advanced of human pursuits, such as that of a scientist, are being transformed. The Center for Computational Learning Systems at Columbia University underlines that robots acting as research assistants have become capable of proposing and testing hypotheses and setting up lab experiments.[5] If this is the state today, what will happen in the next few decades? To be able to imagine and manage purposefully and accurately, it is imperative to understand the potential of the digital age—an age marked by revolutions in the underlying technological foundations.

Technology Foundations of the Digital Age

A complex combination of hardware and software forms the foundation of the digital age. The potential of the digital age depends on the advances in both the hardware and the software. Further, as their combinations advance, radical transformations manifest. Transformation in autonomous mobility is a clear example. Many had ruled out the possibility of robots ever being able to move around intelligently.[6] Indeed, driving or navigating space involves various sensory inputs. However, many digital machines have shattered the wrongly held assumption that senses may not be easily replicated in machines. Robotic vacuum cleaners from iRobot are capable of moving around and even climbing stairs within a house.[7] House-cleaning robots are now widely available, and hardware and software continue to advance rapidly.

The transformative impacts of hardware and software manifest as they enact *binary logic*. Simplistically, binary logic involves thinking and calculating in terms of 1s and 0s—the two binary states. To calculate 2 + 3, you represent the two numbers as a sequence of binary states (0s and 1s). Addition involves toggling the state (0 or 1) in the sequence using binary logic. Advanced software (such as computer-aided design, or CAD) uses the same logic. Such software represents a design, such as a house, as a sequence of binary states (0s and 1s). Transforming the design (e.g., changing the house's colour) involves transforming the sequence of binary states. The ability to enact binary logic depends on the state of the underlying hardware and software; the better they are, the better their ability to store or toggle binary states. When hardware and software enact binary logic better, the transformative potential increases. Not surprisingly, leaders are more successful at digital transformation when they can understand and comprehend the digital foundations. For example, Bill Gates underlines:

> The combination of chips and software show up in a lot of form factors nowadays. The most popular are mobile devices that fit in our pockets. But we've also got that

combination powering our TV sets. We've got it in our cars. It's pretty pervasive at this point. Eventually, we'll get glasses with augmented reality that are adopted broadly, and we'll get robots that are more than simple task-repeaters on production lines.

The sky is really the limit as these chips get more powerful and as the AI software algorithms eventually figure out how to create personal agents that help us get a variety of tasks done like reading and providing advice on scientific discovery. The field still has a long way to go–it's really exciting.[8]

So, to understand the transformative potential, we think a little about the advances in hardware and software.

The Transformative Potential of Hardware

Hardware refers to physical devices and may include servers, computers, mobile phones, sensors, networks, fibre-optic cables and so on. Physically, information (the sequence of 0s and 1s) is stored in tiny hardware units, or transistors. Large physical devices are built to combine many of these basic units. The digital age is powered by advances in hardware. For example, Facebook utilizes large server racks in its data centres (larger than 3,00,000 square feet), spanning across continents, to store massive data for over a billion users. Over the last few years, the digital age's transformative potential has increased due to innovations in hardware. Within organizations, these may be seen in the use of digital subscriber lines (DSLs), and across countries—such as China, India, Chile and South Korea–these advances may manifest in the use of hybrid fibre coax (HFC) cables to offer broadband services (like TV high-speed Internet access). The use of such network-related hardware has transformed connections across individuals, organizations and countries.

Similarly, hardware powering wireless in local loop (WLL), cellular or satellite connections is transforming the potential for individuals to connect and access information. Operating with geosynchronous (GEO) satellites, very small aperture terminals (VSATs) are linking banks in Brazil's remote areas, and using low earth-orbiting (LEO) satellites, global mobile personal communication systems (GMPCS) are providing voice and low-data services.[9] Broadly, large undersea cables, routers and switches have helped connect people across the world. The Internet powers these connections, and over 85 per cent of individuals can access the Internet in the Americas and Europe. Across India, the advent of mobile technologies and providers (such as Jio) has transformed Internet connectivity. Users can now exchange messages, voice, data, documents and more. However, many areas still lack access. Only around 37 per cent of the population has Internet access in Africa (ITU, 2023).[10] Further, much untapped transformative potential will be realized as new hardware, such as new AI-based semiconductor chips, emerges.[11] New types of hardware are emerging, too.

Sensors: The Micro Superheroes

Sensors are a class of emergent hardware technology and they are driving radical digital transformations. They enable the capture of information from across spaces, locations and even human bodies or minds. For example, sensors used in steel furnaces may read temperature during the steel-making process. Similarly, oxygen sensors are used for remote sensing in space, medical and agricultural applications.[12] In law enforcement, sensors are being used in breath analysers to detect alcohol content.[13] Sensor innovations on satellites (NOAA-6) have made it possible to detect gas flares in oil fields on Earth.[14] I am sure many of you have noticed the potential of sensors in your phones,[15] and these transform communication. Ever wonder how a phone switches the screen off when you hold it close to the ear? A proximity sensor comes into action, informing the phone how far your face is to prevent erroneous button clicks when it is held close. Similarly, a location sensor used by a compass in the phone enables GPS functionality and a motion sensor or accelerometer helps discern the phone's orientation, enabling a change of display to landscape or portrait mode. Beyond hardware, exponential advances in software algorithms catalyse the transformative potential of the digital age.

The Transformative Potential of Software

We cannot overstate the advances in algorithms and their transformative impacts over the last fifty years. Google's search is powered by a page-rank algorithm, which is based on a Markov chain model. It rates a web page's importance based on the number of other pages linking to it and iteratively weights it based on the importance of the linking pages. Similarly, Apple's Siri is made feasible through hidden Markov chain models that enable speech recognition. In other domains, algorithms using Bayesian networks help create spam filters. Google's AdSense system is used to customize the placement of ads on web pages[16] and Facebook's FLOW is being used to identify faces in photos and add audio captions for pictures. It forms the foundation for many of the advances at Facebook.[17] Beyond our day-to-day lives, software algorithms are transforming warfare as well. Algorithms coded using Kalman filters enable missile cruises and space travel.

The software algorithms systematically define the logic to create, store, process or transfer binary states (0s or 1s). They do so in a language that is closer to the human mind than to the machine. Simple software often replicates standard human thinking: how we sequence activities (e.g., brush teeth, then take a shower and subsequently have breakfast), select from options (e.g., choose to take an umbrella with us or not) and iterate (e.g., have the alarm go off at 6 every morning). This sequence, selection and iteration often represent the classic control structures at the heart of many early programs and may be prevalent even today. Beyond simple control structures, core complex algorithms are now commonplace. The use of AI to find patterns in mass data is becoming mainstream, and contemporary cognitive computing systems (such as IBM's Watson) exhibit the transformative potential of modern algorithms.[18] Similarly, algorithms for information exchange over communication channels may leverage Shannon's theory of communications.[19] Generally, growth in algorithms is driven by the growing focus on existing and new scientific advances in algorithms: regression analysis, data mining, supervised or unsupervised deep learning, classification, support vector machines, optimization, clustering and probability estimation.[20]

Not surprisingly, modern software applications are very effective at processing binary logic. For example, the computer programming language Mathematica has been used to model and assess the behaviours of approximately 50,000 engineering components, such as those used in airplanes.[21] Software has advanced so much that it superseded human abilities many years ago. As is widely known, in 1997, Deep Blue defeated Garry Kasparov. This was no mean feat, as chess was widely considered a game that required deep human intelligence. Going beyond chess, AlphaGo, built using Google's DeepMind, beat a human, Ke Jie, the number one player, in the Go game.[22] Go was widely regarded as multiple times more complex than chess. However, advances in software are overcoming the human mind. In turn, they are transforming the potential to enhance human lives. Even before the advent of ChatGPT, by 2009, AI-based natural language processors (such as Apple's Siri) provided low-cost technology-enabled secretarial functions, such as voice-based email, phone calls, search and appointment management. The evolution of software is only increasing with growth in the areas of optimal computing, biological (DNA) computing, quantum computing or algorithm design, such as evolutionary algorithms (EAs), swarm intelligence (SI) algorithms and artificial neural networks (ANN) algorithms.[23] In general, algorithms are taking centre stage, catalysing rapid advances in software and enhancing the transformative potential of the digital age.

However, despite these tremendous advances, there is a problem: many digital transformations are not successful.

Transformative Paradox

A paradox has emerged, as despite the deep transformative potential, many have raised a red flag about digital transformations. First, many are worried about digital transformation increasing the threats of social inequity, the creation of large monopolies, mass unemployment and job losses. In the US, inequity is increasing, as reports have highlighted that a small group (typically defined as the top 1 per cent) has benefited disproportionately.[24] Widespread inequity may be dangerous. Concerns about inequity in well-being have led to upheavals across countries, such as Egypt and Britain.[25] This concern was echoed in the US presidential elections in 2016 as well. Former technology entrepreneur and participant in the Democratic presidential primaries, Andrew Yang, linked this unemployment with the advanced digital transformations of work.[26] Specifically, people worry about the systematic transformation of the world, whereby machines become the predominant workers. Indeed, digital transformations are leveraging the fact that technologies replicate, or even surpass, most human skills and labour-related tasks. In most cases, it is impossible for humans to compete with these technologies. You may think about ChatGPT and how many of the tasks it carries out, such as formatting references, quick creation of summary tables and similar others, are going to make some jobs redundant. Sure, people will become more productive, but it will also mean fewer employees are needed to do the same job. Unsurprisingly, even the most ardent technology enthusiasts are raising concerns.

The conclusion that digital technologies *may* (potentially) be associated with loss of work and well-being is dangerous; at the very least, it is disheartening. Unfortunately, it can become real! Erik Brynjolfsson at MIT analysed data to show that there is a decoupling between jobs and productivity. In simple terms, it is now feasible to organize effectively without employing more people. That is, we may reduce the number of people working in an organization to create more advanced products and services. Erik summarizes the argument by suggesting that the most valuable organizations of the twentieth century, such as Ford, employed many more people at their peak than is being done by leading organizations of the twenty-first century (such as Facebook).[27] If the finding is pervasive, it may mean a substantial reduction, if not a near end, to an age whereby human work led them to prosperity.

Second, many have raised concerns about rogue actors using these technologies to do illegal or immoral things. Even the leaders in the domain of digital transformations have vocalized their fears about advanced technologies, such as AI. Speaking at the George W. Bush Presidential Center's Forum on Leadership, Amazon founder Jeff Bezos comments that current AI technologies are adequate to create autonomous weapons and '. . . these weapons, some of the ideas that people have for these weapons, are very scary.'[28] Finally, there is a real concern at the personal and relational levels. The modern era of technology is transforming human relationships and social interactions (e.g., through social networks such as Instagram). Alongside the positives of using these technologies, there are concerns related to adverse outcomes, such as cellphone addiction, depression, echo-chambers, viral fake news and other such phenomena.

In summary, despite its tremendous potential, the overall sentiment about digital transformation is not universally positive. I call this the *transformative paradox*. Resolving the paradox quickly is crucial because advances in hardware and software are exploding. Colvin underlines that if computing power doubles annually, it will be a million times more than today *in the next two decades*. (You may calculate what doubling for twenty years means!)[29] Now, compare this with the growth in traditional ways of life. You will be shocked. There is only a 100-fold difference between the speed of a human walking and the speed of a jet plane.[30] Think: a) Today's smartphones compute better than the most powerful computer in the early 1980s, and what will the mobile in human hands be like in the next twenty or even fifty years? b) Approximately every two years, the data created in the world is more than the cumulative data created up to that time, so what information about our lives will *not* be stored and computed digitally in the next five or twenty years? c) Developments in natural language processing (NLP) led to Apple's Siri and Amazon's Alexa in 2009 and ChatGPT in 2023. Many of these are transforming white-collar jobs. So, what will life and work look like in the next twenty years? Are you excited about the potential of the digital age? While several exciting opportunities lie ahead, many also anticipate new challenges.

Successful Digital Transformation: The Purpose

This book resolves the transformative paradox by outlining how to lead successful digital transformations. While the dark side of digital transformation is real and needs to be worried about, the positives of digital transformation are exponentially greater than the negatives. The primary reason for the paradox is the difference in the purpose of digital transformation. The purpose shapes how we transform our lives through technologies such as Google Search or Maps, UPI-based payment apps (e.g., PhonePe, Paytm and Google Pay) and many others. Different constituents—entrepreneurs, social voices, political leaders, venture capitalists and so on—think about the purpose of digital transformation differently. People differ in their perceptions of how technology can transform their (and others') lives. A constituent with Purpose (A) may be at odds with another constituent's Purpose (B) for digital transformation (See Figure 1). What is the idea of purpose, and how may we pursue a purpose that leads to our own and others' well-being?

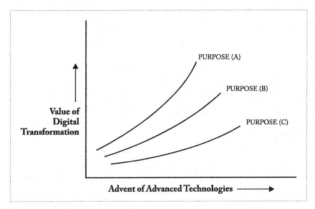

Figure 1: Transformative Paradox Differences in Purpose of Digital Transformation

I argue that this question is crucial for successful transformation—as a society, individual, family, nation, business or any other entity. Digital organizations are everywhere and they are impacting everyone. Air is probably the only element that is more pervasive. However, the purpose of harnessing their potential is hard to master. Over the years, I have found that understanding digital transformation is not solely about knowing and being familiar with technology. Rather, it is about having the wisdom to leverage them *purposefully*. I do not think I can emphasize the importance of wisdom to pursue a purpose more cogently than the way Stephen Hawking[31] reasoned: 'Our future is a race between the growing power of our technology and the wisdom with which we use it.'[32]

I explain and outline the methods that lead to successful digital transformations. Specifically, I outline a model of purpose that underlines successful digital transformations.

Conclusion: Treatise on Digital Transformation

Why do we need to study digital transformation? There is a lack of clarity on what digital transformation is. While people have heard of popular technologies such as AI, NLP, the Internet of Things (IoT), Deep Learning, etc., digital transformation remains a nebulous concept for many. What kind of imagery does the term 'digital transformation' evoke for you? Does it energize you to act, fill you with optimism for the future and make you thrilled to think about the endless possibilities for yourself or your children? Most of the people I interact with—policymakers, business leaders and students—have some answers, but they are rarely deep and clear. When I began researching the topic in the early 2000s, it became apparent to me that digital transformation is not a trivial process. One may not rely on a quick way to adopt technologies and make them work. However, what is certain is that there is a deep, purposeful and ongoing process. The technology-led transformational journey of human society started approximately 5500 years ago.

What was life like before 3500 BCE? Imagine how we lived when the wheel—a technology so pervasive today—did not exist even in its basic form.[33] After its advent, the wheel transformed transport (of people and goods). Subsequently, many eras, such as the Stone Age, the Bronze Age and the Iron Age, leveraged the invention and built new technologies that transformed travel. The Industrial Revolution led to more advanced technologies for transport. Humans could travel much faster and more comfortably with locomotives made using mechanical or electrical machines. The second machine age, as it is sometimes called, is further transforming travel through autonomous cars.[34] Generally, lives are being transformed through digital innovations—digitized dishwashers, software-embedded cars, washing machines, mobile phones, autopilot airplanes or language models such as ChatGPT. Beyond transport, the ongoing digital age is transforming: a) work (e.g., we make plans using digital boards and execute them using robotic process automation, or RPA[35]), b) wars (e.g., missile shields protect from missiles coming from across our borders) and c) healthcare (e.g., we expect to live longer and healthier through digitally-driven medical devices, apps and healthcare models (e.g., patient-engaging care models).[36] Their impacts on our lives and organizations are deep. Indeed, as Douglas Engelbart underlines, 'The digital revolution is far more significant than the invention of writing or even of printing.'[37] That is, the transformation is not only pervasive but also purposeful, with a deep historical continuity. One has to synchronize digital transformation with the broader purpose—the continuous evolution of humankind.

These digital transformations are radically changing the *purpose* of our lives. As I will later elaborate, life's *purpose* is to evolve—academically, scientifically, materially, religiously, intellectually and so on. As you delve into the book, consider how the digital transformation of the world is shaping your life's purpose. Ask yourself: Are my endeavours to digitally transform in sync with my purpose—the human purpose? To help guide you in the endeavour, the book presents a model of purpose that drives successful digital transformations. The model emphasizes that digital transformation is more than simply acquiring technology. Instead, it requires making a deep change. This is because digital transformation fundamentally alters our lives. Pause for a moment and think about how these digital technologies are transforming our lives, making us rethink the basic principles underlying our behaviours and thinking.

That is, go beyond thinking about digital transformation and use the book to contemplate the fundamental questions surrounding life and existence. As I have observed and will explain, there are deep linkages between life and digital transformation. Digital transformation succeeds when it is done with the wisdom and purpose to enhance human happiness, well-being, peace and continued innovation. In this book, I emphasize that understanding the *purpose* bears the wisdom necessary to tackle life's complex questions. To do so, one has to unravel the linkages between the two: the abilities of digital technologies and human organization. The book builds a model to enact purpose by answering the three questions: What drives successful digital transformations (Part 2)? How does one plan and lead a successful digital transformation (Part 3)? And, why should we think about digitally transforming (Part 4)? I start next by proposing an organizational approach to scientifically unravel and master the purpose of digital transformation.

Part 1
Digital Transformation

2

Digital Transformation of an Organization

The Scientific Foundations

'Science is a way of thinking much more than it is a body of knowledge. Its goal is to find out how the world works, to seek what regularities there may be, to penetrate to the connections of things—from subnuclear particles, which may be the constituents of all matter, to living organisms, the human social community, and thence to the cosmos as a whole.'

—Carl Sagan[1]

Digital transformation is a transformation of the organization. Digital technologies have transformed a) how we organize our finances by using digital payment systems (Google Pay, Paytm and other UPI-based apps); b) family life through smart home technologies; c) community connects through shared housing and co-living spaces (e.g., Airbnb); d) work environments through the co-working spaces where digital technologies facilitate the ability to schedule and reserve desk spaces; e) collective learning through hackathons, open innovations or collaborative platforms (e.g., Stackexchange.com);[2] f) foreign explorations and interactions through online travel websites, blogs and virtual reality experiences. These transformations may seem distinct and different, but they are not! Rather, they are driven by the digital reorganization, or *digital transformation of our organization.*

Historically, whenever new technology has come to the fore, a clear purpose has been the driving force behind successful transformations. Consider the transformative impacts of electricity and railroads on organizations. The advent of electricity in the US transformed factories.[3] In the older generation of factories, machines were powered by steam or water pressure. A complex combination of pulleys and gears helped trigger the machine's operations. However, the advent of electricity brought about a transformation in the factories. Nick Carr, editor-at-large, *Harvard Business Review*, argues that the purpose of transformation across many factories was to replace the source of power, while still using the system of pulleys and gears throughout the factory. However, some remarkably successful transformations pursued a different purpose. Thinking deeply, some entrepreneurs reasoned that the advent of electricity calls for a factory redesign. Unlike water or steam, it was easy to distribute and bring power directly to workstations. So, they rewired their plants and installed power sources next to the machines. These made the operations nimble and efficient. Soon, the productivity of these organizations soared, giving them a competitive advantage over their peers.

A similar difference in the purpose of organizational transformation was evident after the advent of railroads in the Americas. When the United States sincerely went about laying down its rail lines in the mid-1800s, organizations found an opportunity to transform. Nick Carr argued that steamships, while capable of transporting goods over long distances, primarily carried raw materials.[4] The advent of railroads changed American industry's transformation, as organizations could now transport finished products inland. Small towns became accessible. So, many organizations transformed purposefully to build large-scale mass-production factories. Railroads transported the goods to smaller cities and towns, giving consumerism a boost. Indeed, throughout the history of civilization, many technologies and related tools, such as the wheel, carts, automobiles, fire and industrial machines, among others, have transformed the purpose of organizations. Digital technologies are doing the same.

Digital technologies have brought products to individual homes, influencing the way kids learn, how consumers shop, how they schedule medical appointments, communicate with friends and relatives and so on. A purposeful transformation of the organization is crucial for success. I teach the idea of purpose to many at the Indian Institute of Management Ahmedabad (IIMA), with academics and practitioners attending from India and abroad. Teachers and professors are now using cutting-edge technologies (such as advanced server farm computers, robots, online learning management systems, or LMS, and so on) to do their research, deliver lectures and grade exams. A purposeful approach to organizing academics improves it fundamentally. To elaborate, I often discuss the role of AI in education. Advanced AI-based technologies for grading school students enable teachers to find patterns in learning deficiencies. These patterns are not noticeable to the common eye. For instance, hidden patterns are revealed when teachers use AI to analyse student mistakes across courses.[5] Through such analysis, one may find that a student is making mistakes across courses as they do not know one specific concept (say, quadratic equations). Such findings are difficult to unravel in a traditional learning environment. A purposeful digital transformation ensures that such algorithms aid student learning and are not a source of abject awe, confusion or anxiety.

To systematically understand the model of *purpose* for transforming digitally, we begin by using the organizational approach, rethinking what an organization is.

Rethinking Organization

The organization is a pervasive phenomenon on this planet. Organization is an act of living beings. All organizations make the world work. One thinks, understands and participates in various organizations since birth. A newborn child, when being fed, is participating in an organization between himself or herself and his or her mother (often) or another caregiver. Over time, as the child grows up, he or she continues to think about, understand and participate in organizations—school, family, playground, swimming classes, birthday parties and so on. As the brain matures, one thinks about these organizations from different perspectives. The understanding of organizations evolves, as it is feasible to have multiple views about organizations due to a more evolved brain. We try to understand organizations as we make sense of the world. We may ask questions such as: How does the tax department operate? How does Google work? How does the new Tesla function? How do engineering, medicine or management admissions work? Further, we also create organizations. Digital technologies are key to organizing today. However, what is an organization? The advent of new technologies transformed the idea of the organization.

For example, the advent of the Industrial Age led thinkers to debate the organization. Theories were put forth, and a noteworthy view underlined that the organization is best understood using a production function approach. This approach flourished in the domain of microeconomics. Greater productivity among workers was one main outcome of this approach. The organization took inputs and processed them to create outputs. As an example, think about a flour mill. The firm takes as inputs capital (to buy the wheat, mill, etc.), land (where the flour mill is housed) and employees (that operate the mill, order grains, pack the flour, etc.). The production function approach emphasized production as the key purpose of the firm. The firm could succeed at its purpose if it continually thought about enhancing inputs—land, labour and capital—to influence the output and efficiency of inputs. For instance, new technologies could increase worker efficiency by enabling them to accomplish more tasks in the given time. Further, one may continuously think about changing the combination of inputs to transform and create a better organization.

Being a dominant paradigm in the Industrial Age, the production function perspective was initially used to transform organizations digitally. Early studies using the approach assessed how investments in digital technologies enhanced an organization's productivity. Specifically, various economics and information systems researchers assessed if the inclusion of IT (in certain amounts, with certain combinations of land or labour as inputs) makes the organization more productive. Some success was realized through this research. Indeed, technologies of the previous era, such as expert systems or decision support systems,[6] office productivity suites (Microsoft's Office or Lotus Notes) and so on, were production-oriented and meant to enhance the workers' productivity. ERP, CRM and other enterprise systems did the same. However, inconsistencies about the impacts of IT on the firm's performance emerged. The concern was so great that Nobel laureate Robert

Solow quipped about it, leading to the popular term—productivity paradox.[7] The paradox underlined that IT was everywhere in organizations, but it was not contributing to productivity.[8] While the paradox was resolved, related research and thinking revealed that digital transformation is a highly creative endeavour.

Managers must pursue a creative purpose to leverage more complex technological innovations. One clear example is the use of AI in many significant present-day technologies. These require an organization to innovate. That is, a purpose focused on innovation is crucial for success and may not be confused with the invention (say, of the technology).[9] As an example, when we think about a food-ordering app (like Swiggy or Zomato), the technologies—Java or Python, Ruby on Rails, Node.js, recommendation engines, payment gateways, map APIs and many others that may have been used—were not fundamentally new inventions.[10] However, these apps innovated as they digitally transformed the food industry, leveraging technology in innovative ways. They collaborated with restaurants and other food providers, and they innovated in marketing to create potential customers' expectations and interest. They also put together innovative logistic operations to deliver on time. As a result, the transformation of food delivery became effective. Similar organizational innovations drive e-commerce platforms (such as Ajio, Amazon and Flipkart), ride-sharing apps (such as Ola and Uber) and e-learning platforms (e.g., Swayam, Coursera and Udacity).[11] To be able to purposefully innovate, an organization has to rethink itself. I use a scientific approach to define the organization and unravel the purpose of digital transformation.

Organization: A Scientific Perspective

An organization is a complex entity that may be best understood using a scientific approach. To do so, it is imperative to delve a little deeper into the question: What is the scientific approach?[12] Science is best represented in the form of theories and models. These help us understand concepts and entities, as well as shape our thinking and imagination. Science offers a perspective, as the American theoretical physicist Brian Greene underlines, 'Science is a way of life. Science is a perspective. Science is the process that takes us from confusion to understanding in a manner that's precise, predictive and reliable.'[13]

Philosophically, science perceives (or imagines) reality. Indeed, it is well known that scientists require imagination to build theories. World-renowned thinker and professor of organizational science, Karl E. Weick[14], defines disciplined imagination as a process to propose new theories. He compares the process of theory building to an evolutionary process, suggesting that good theories emerge as a result of alternate hypotheses. As new theories emerge, reality is seen differently.

Consider how the world 'saw' (thought about) light. For a long time, scientists confirmed that light is made of particles. The particulate theory offered a lens to observe the subatomic world as a well-organized set of moving particles. However, the lens was challenged. An alternate view of the light proposed it to be a wave. This view was also found inaccurate when an alternative lens (theory) highlighted that light is sometimes a particle and at other times a wave. This manifests as Heisenberg's uncertainty principle. The principle states that particle or wave nature manifests depending on whether one observes it or not. This view was held by the likes of Albert Einstein. He argued that quantum-scale objects may not exclusively possess the properties of either particles or waves.

In general, science emphasizes how it shapes the current perspective humans have about a phenomenon. Specifically, the prevailing theory shapes this perspective. A new theory changes the perspective, even when the empirical facts may remain almost the same. Regardless, new scientific theories shift perspectives. As argued by the Nobel laureate, Sir William Bragg: 'The important thing in science is not so much to obtain new facts as to discover new ways of thinking about them.'[15]

Different theories create different perspectives because they all offer a unique lens into reality. This has huge implications for understanding digital transformation. Specifically, I use this scientific approach to make sense of an organization. I use multiple scientific lenses (theories) to unravel multiple facets of an organization. Indeed, organizing is a complex endeavour. So, defining an organization is challenging, and the organization is often too complex to be understood completely from any one scientific perspective. Any one facet of an organization offers limited understanding. Not surprisingly, scientists unravel the principles of the organization using different theories. Each perspective shows one facet of an organization. The process of using multiple lenses is like blindfolded wo(men) 'seeing' an elephant by touching a part of it—a knee, a tail, a trunk or a tusk—to assess what it is. Based on the part they touch, each 'sees' the elephant differently. That is, based on what image the part conjures for the woman or man, the elephant is a very distinct entity. One person may perceive the elephant's trunk as a swing, another may perceive its leg as a pillar and so on. However, when we combine, discuss and reflect on all of these descriptions, we get a more accurate picture of the elephant.

To summarize, because each lens (scientific theory) may offer a view of only one organizational facet, I would use multiple lenses to *see* the organization from different perspectives, outlining the ways to digitally transform. Specifically, throughout the book, I will conceptualize the organization using some of the following scientific lenses:

- Capabilities Lens: The routine (efficiency)-based view of the organization (Chapters 3 and 4)
- Computational Lens: The information processing view of the organization (Chapter 6)
- Work System Lens: Organization and work (Chapter 7)
- Decision-Rights Lens: Organization as a bundle of decision-rights (Chapter 8)

- Value Lens: Organization as a value-creating entity (Chapter 9)
- Learning Lens: Organization balancing exploration and exploitation (Chapter 13)

When an organization has to be analysed and managed from many different perspectives, how do we get a holistic picture? The *purpose* integrates the organizational digital transformation.

Purpose and Scientific Forces

Purpose offers a latent force for the digital transformation of the organization. A scientific approach to a phenomenon often relies on identifying latent, yet potent, forces. It is the hallmark of science to unravel *forces* underlying phenomena. For example, physicists talk about the force of gravity that explains a wide range of phenomena, from the movement of celestial objects and satellites to the movement of cars on an F1 formula car-racing circuit. Beyond the force of gravity, physicists have also found latent scientific forces that guide intelligence—inquiry, thinking and research. Such a force enables physicists to unravel natural laws. Nobel laureate Murray Gell-Mann argues for *symmetry* and *beauty* as such a force. While beauty may seem to be an anachronous concept for physics, it helps unravel natural laws. Gell-Mann outlines that the beauty of an equation, due to its symmetric formulation, makes it reveal the natural truth. He compares various equations used to describe natural reality and underlines how common symmetry and simplification make them reveal more of the natural world.[16] Similarly, chemists proposed a periodic table based on a symmetric placement of atomic particles and helped organize the elements in the universe, even predicting the presence of elements that were not discovered at the time. The force unleashed by symmetry has been strong, latent and hidden, but very potent.

For digital transformation, the purpose is such a force. Identifying a force is the hallmark of scientific inquiry. I underline that purpose is a force that drives high-performing digital transformations. Specifically, it helps unravel a pattern, a trend or a reason that catalyses the evolution of life, organization and digitization. The advent of COVID-19 triggering online shopping is an example of how humans enacted transformation purposefully. Generally, the force of purpose guides successful transformations, and I unravel this force by studying the digital transformation of an organization

Conclusion

I argue that digital transformation is a concept that may not be mastered using any existing organizational theory. Over the years, I have researched many digital transformations, analysing them from different theoretical perspectives: economics, physics, computer science, organizational theory, strategy, information systems and computational neuroscience. Many of my findings and insights, from various research studies, consulting assignments and industry executive training, have helped me unravel different facets of digital transformation. Specifically, these underline the need to apply a rigorous scientific approach to understanding organization. Using a scientific approach, the book offers a treatise on the digital transformation of *human organization*s.

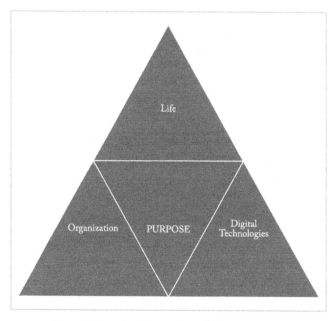

Figure 2: Purpose Driving Digital Transformation

I would use multiple different lenses (theories and models) to study an organization and its digital transformation. Collectively, when we use different theories, we overcome our limited ability to comprehend the underlying *purpose*. This purpose underlying life and organization guides digital transformation, and clarity about the purpose is required to achieve the creative potential of digital transformation. To reiterate, *purpose* is a natural force that drives digital transformation. I outline how it drives the intricate linkages between life, organizations and digital technologies (Figure 2). I unravel *the purpose* by answering three questions: what, how and why do organizations transform digitally. Specifically, I explain and define three facets of purpose:

1. The instrumental purpose that answers: What is the goal of an organization's digital transformation?
2. The operational purpose that answers: How should the organization transform digitally?
3. The existential purpose that answers: Why should the organization digitally transform?

These purposes—instrumental, operational and existential—comprise the force that drives digital transformation. These three are intricately intertwined. In this book, I outline a *model of purpose* for digital transformation. Beyond helping understand the organization, the model helps individuals unleash the force (the vision, the optimism and the energy) required to harness the transformative potential of digital technologies.

Part 2
The Instrumental Purpose of Digital Transformation

3

The Instrumental Purpose of an Organization's Digital Transformation

'Indeed, in India, digitization is quickly becoming a core capability as digital public infrastructure is being widely developed enhancing capabilities of individuals, say through Digital India initiative.

'One of the most exciting success stories has been the digitization of small businesses. Just four years ago, only one-third of all small businesses in India had an online presence. Today, 26 million SMBs are now discoverable on Search and Maps, driving connections with more than 150 million users every month. What's more, small merchants across the country are now equipped to accept digital payments. This has made it

possible for more small businesses to become part of the formal economy, and it improves their access to credit.'
—Sundar Pichai, CEO,
Google and Alphabet Inc.[1]

What is the instrumental purpose of digital transformation? That is, what should an organization strive to achieve through digital transformation? The goal of all organizations is greater performance. Indeed, all organizations aspire (and rightly so) to perform better.[2] Also, organizations leveraging the power of digital technologies have outperformed others in various domains, especially in the corporate world. For example, in the US markets, various acronyms such as FANG (for Facebook, Amazon, Netflix and Google) have become popular to outline elite digital organizations. These companies are beating traditional companies by a long margin. Bank of America's data reveals that only seven stocks—Microsoft (MSFT), Alphabet (GOOG), Apple (AAPL), Amazon (AMZN), Tesla (TSLA), Nvidia (NVDA) and Meta Platforms (META)—made up close to $11 trillion in market value in 2023.[3] In the corporate world, the inability to perform gives competitors the space to grow. Thereby, high-performing competitors take over in a short time. In North America, the fall of Blockbuster is an example that underlines the perils of not leveraging digital technologies effectively and losing to a nimble competitor (in this case, Netflix).

Not surprisingly, digital technologies are now being used widely across organizations, beyond the ones in high-technology industries and the corporate world. For example,

a) to adjudicate cases quickly and accurately, courts are using digital case-management and filing systems and video conferencing for remote hearings, b) to enhance student learning outcomes and research output, universities are using virtual classrooms, learning management systems and research databases, c) to enhance patient health outcomes, hospitals are using electronic medical records, telemedicine and medical imaging systems, d) to enhance research outputs, research organizations are using high-performance computing, data visualization tools and machine learning algorithms and e) to enhance customer service performance, financial organizations are using mobile payment apps, automated trading systems and financial fraud detecting AI technologies. However, is adopting advanced digital technologies sufficient to realize high performance? The answer is overwhelmingly 'no'. The organizations that thrive (and survive) are the ones that pursue a clear instrumental purpose while digitally transforming. What is the instrumental purpose that high-performing organizations pursue to leverage these digital technologies?

Answering this requires assessing what leads an organization to perform better. And defining this has not been easy. The works of noted scholars offer some insights.

Models of Organizational Performance: Economic vs Strategic Models

Organizational performance has been well studied. Scholars underline models predicting high performance.

Very broadly, these models may fall into two categories: position-based versus efficiency-based, outlining two markedly different instrumental goals.[4] First, the positioning perspective argues that the performance of an organization is dependent on its position in the industry. Widespread teaching of the perspective in management and business schools led to its popularity and status growth. The perspective developed primarily in the works of Michael E. Porter (1980).[5] To understand the perspective, think about a flour mill. How may its position enhance performance? A flour mill is not just profitable because it can grind input into output, but its performance depends on how much it can bargain with suppliers, command a superior price from customers and ward off the threat from substitutes—such as packaged flour—and new entrants—other flour mills that intend to set up operations to target the flour mill's customers. These are the five factors that Porter's positioning-based model identifies as drivers of performance. The positioning perspective resonated in the industrial age, as organizing became a widespread activity. With the advent of digital technologies, there is a growing focus on internal efficiencies rather than taking industry positions. There is a shift in thinking about what leads to superior performance in the digital age.

Performance in the Digital Age

In the managerial hallways, it is evident that the advent of digital technologies has led to a move beyond positioning-based logic. Efficiency-based models are gaining ground. These models underline a focus on internal firm dynamics. This is seen most clearly from two perspectives: the resource-based view[6] and the capabilities perspective.[7] The two highlight that superior performance is a result of the possession of valuable, rare, inimitable and non-substitutable resources or the development of high-performance capabilities, respectively. Considering the previous example of the flour mill. The mill can realize superior performance through internal excellence, achieved by hiring highly skilled workers or through rare machinery (resources) that could grind the grain into flour in a manner that competitors could not (resource-based approach). Alternatively, the flour mill could build advanced capabilities, such as processes to deliver customer orders at home in a short time.

Indeed, historically, various innovations have enhanced the capabilities of human organizations. Radiocarbon dating presents evidence of how swords, combs or pots enhanced human abilities. Similarly, the invention of the wheel revolutionized how humans travelled and transported goods, and the spear changed how humans defended themselves and what they ate.[8] Also, there is evidence that in England, heart-shaped hand axes were used to cut meat and create food. The use of clay seals with simple geometric patterns and figures of animals and people in the Middle East during 6000 BCE indicates organizational processes enabling productive work. There are numerous other examples of technologies and artefacts enhancing human capabilities, such as a) tablets that organized writing endeavours in Syria during the fourth millennium, b) razor-sharp blades that helped catalyse hunting in North America between 8000–1000 BCE and c) roads that enabled travel for state business in the Persian Empire during 550–331 BCE.[9] Beyond these, evidence indicates that human organizational capabilities could catalyse complex procedures. For example, in 400 BCE, Peru, medical artefacts enabled practitioners to perform trephination, which involved the surgeon removing a part of the skull to alleviate pressures due to falls or blows and then closing the surgical hole in the head.[10]

Contemporaneously, such advanced organizational capabilities are digital. For example, AI-driven capabilities promise to transform healthcare outcomes, especially in poor countries. Bill Gates outlines the potential of these digital capabilities:

> . . . many people in those countries never get to see a doctor, and AIs will help the health workers they do see be more productive. (The effort to develop AI-powered ultrasound machines that can be used with minimal training is a great example of this.) AIs will even give patients the ability to do basic triage, get advice about how to deal with health problems, and decide whether they need to seek treatment.[11]

To summarize, it is now increasingly being realized that *the instrumental purpose of digital transformation is building digital capabilities.* The theoretical foundations of capability logic have a long history. It has its academic and theoretical basis in the view of the firm that emphasizes efficient administration, and the idea of 'routines' proposed by Nelson and Winter in 1982.[12] However, the term 'dynamic capabilities' was most famously attributed to the research by David Teece and colleagues published in 1997. Specifically, David Teece, Gary Pisano and Amy Shuen formalized the Capabilities Model of organizational performance, arguing for a focus on building dynamic capabilities.[13] They argue that high performance arises when organizations develop the ability to adjust internal and external competencies to respond to environmental demands. Thereby, the instrumental purpose of digital transformation underlines leveraging technology to build dynamic capabilities that help an organization adapt to changing circumstances.

The instrumental purpose underlines a capability-performance hypothesis. This hypothesis is used to assess which digital capabilities enhance performance. Generally, digital capabilities have been found to enhance the performance of building new products,[14] managing knowledge[15] and serving customers.[16] I have been fortunate to conduct some well-known research studies testing this hypothesis, along with my co-authors. In these research studies, we have unravelled new ways to build and harness digital capabilities for creating superior organizational performance. For example, we unravelled ways to leverage IT for agility in the supply-chain and demand-management domains. Similarly, in the customer service domain, we unravelled two key capabilities that enhance customer satisfaction: customer orientation and customer response capabilities. This was a notable contribution. Specifically, outlining the role of IT in the customer service domain, we unravelled how managers need to ensure the quality of information available across a bank's branches, as these influence the two important (and hitherto unknown) customer service capabilities.[17] Similarly, in other related research on the topic, along with my co-authors, I differentiated the potential and realized knowledge management capabilities. Specifically, we identified how managers may integrate IT systems to build capabilities to acquire, assimilate, transform and utilize knowledge.[18]

Even outside the organization, such digital capabilities are revitalizing organizational performance. For example, Bill Gates underlines that the use of AI is helping farmers develop better seeds and better plantation strategies, as well as drugs and vaccines for livestock.[19] Similarly, Gini Rometty, former CEO of IBM, explained how the digital initiative CRUSH (Criminal Reduction Utilizing Statistical History) enhanced the performance of the Memphis Police Department. Specifically, it created the capability to reduce rapes by 30 per cent as it moved the pay phones indoors after finding a correlation between rapes and pay phone location (outdoors).[20] In summary, high performance results when an organization pursues an instrumental goal to build advanced digital capabilities.

Instrumental Purpose of Organization: High-Performance Digital Capabilities

At large, digital capabilities help an organization perform exponentially better. These capabilities are now everywhere. Amazon has developed the capability to ship the products before you place an order. Defined as anticipatory shipping, the concept was described in their patent and involves the retailer shipping products without specifying the final address. The digital capability for anticipation is built using data from various sources, such as customer surveys, customers' wish lists or browsing and shopping patterns. The digital capability is valuable for Amazon. It helps the company find how profitable it is to keep products in transit, say, on trucks that are in the vicinity of the customer who eventually demands it. It quickly delivers when the demand materializes. By meeting the demand quickly, Amazon creates more loyal customers and enhances repeat purchases, adding to its bottom line. Building such a capability without using digital technologies is implausible, if not impossible. Many other digital capabilities are now creating a huge value. Some of the examples include the ability of online insurance companies to issue and send a policy within minutes of a complete application, the ability of online retailers to share recommendations based on recent and past purchases, the ability of government portals to accept and submit tax refunds within days (not weeks), the ability of social networking platforms and instant messaging applications to enable communications amongst multiple people across the globe and so on.

In their *Harvard Business Review* article, professors Erik and Andrew talk about how big data is helping organizations create advanced digital capabilities. Two cases they discuss are worth mentioning. The first is that of Sears Holdings. It used a Hadoop cluster to collect and analyse petabytes of data, at a very low cost, from all its different brands. It then directly analysed the cluster's data. This shortened the time to launch promotions from eight weeks to one. The second example is the use of the service Right ETA by PASSUR Aerospace, an aviation company. PASSUR Aerospace provides decision-support technologies that use big data, such as publicly available weather data, flight schedules and plane-location feeds from passive radar station networks near airports, to estimate the arrival time of a flight. Millions of dollars are saved,[21] if the deviation (in minutes) between actual and stimulated arrival times is minimized. These performance gains arise due to savings in operational costs that arise because the staff is still idle or stuck when the flight lands.[22]

At large, it has become apparent in the last few years that organizations perform well when they build *digital capabilities*. Leading retailers have developed and deployed capabilities to recommend what other products customers may want to buy by analysing their and others' purchasing behaviours. These recommendation algorithms rely on the ability of the firm to get the data, combine it with advanced scientific algorithms and integrate an understanding of marketing and consumer behaviour. This involves managing complex interactions between different processes, departments, technologies and data. Generally, the emphasis on capabilities challenges the notion that performance manifests due to an organization's position. Also, firms create capabilities that may not be known in advance. Often, human creativity and ingenuity shape the capabilities built by an organization. That is, organizing is the process of creatively acquiring and using resources through activities that may make the firm capable of performing valuable tasks. To summarize, digital capabilities are abilities built using digital technologies, and these lead to better performance.

Conclusion: The Organizational vs Tools View

It is imperative to emphasize that the purpose of digital transformation is to enhance organizational digital capability. Indeed, designers create all technologies with inherent capabilities that manifest in their features and affordances. These affordances represent the designer's perspective on what the technology capabilities and use cases are likely to be. The technology's capabilities are often well advertised and highlighted. For example, you would have often heard about resource pooling, self-service, redundancy and other such functionalities of cloud services. These functionalities manifest the designer's intent. The designer's intent considers the likely usage of the technology and its potential sufficiency in certain contexts. Relying on these technology functions underlines the technological (or engineering) approach. However, digital transformation is best championed using an organizational approach.

That is, in most academic research, the two views are prominent and often positioned as orthogonal (opposite) to each other. The first is often defined as a *tool's view* and visualizes digital technologies in terms of their engineered purpose—their technological capabilities. The second view is defined as the *organizational view* and underlines how digital technologies are structured within the organizations as users adapt and transform technologies to build organizational digital capabilities. Different technology capabilities are used differently because of the managerial creativity and dexterity.

For example, blockchain is a technology that uses a shared and immutable ledger of transactions, so the data is openly visible to multiple parties. However, there are many different organizational models for creating blockchains. IBM's Hyperledger, Ethereum and others operationalize blockchain models differently as they pursue different purposes and approaches. IBM's Hyperledger enables a permissioned and private network only, while Ethereum operationalizes a public or private configuration. Similarly, Hyperledger does not have native built-in cryptocurrencies underlying transactions, while Ethereum has a built-in token. Generally, digital technologies have raw capabilities when they are created or engineered. Digital transformation's instrumental purpose is to innovatively morph these digital technologies into digital organizational capabilities in line with the goals of the organization (see Figure 3). Senior digital leaders underline how organizations will transform themselves to leverage these technology capabilities. For example, Bill Gates argues that industries will be reoriented around AI and that organizations will realize performance only by learning how to use AI.[23]

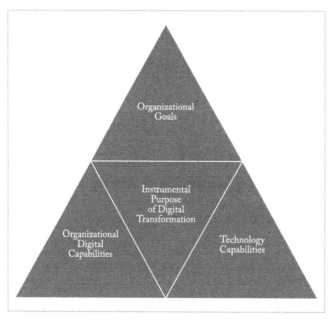

Figure 3: The Instrumental Purpose of Digital Transformation

The instrumental purpose helps transform and align the technology capabilities to create organizational digital capabilities. To be precise, the instrumental purpose underlines the organizational view. This view outlines that for any organization its goals are paramount, not the technology designer's intent (see Figure 3). However, technology capabilities offer an exciting opportunity to reorganize.

Building a capability, not buying it, is the key to its high performance. Indeed, this view drew attention to the value of internal firm dynamics. Leading organizations such as Amazon have shown how advanced capabilities are created using modern technologies such as AI, robots, drones and NLP. In some of the works discussed above, I have outlined my research that has unravelled digital capabilities created in operations management (for managing knowledge) and in financial organizations (for servicing customers).[24] Creating digital capabilities is complex and may require managers to master ways to relearn and reconfigure their processes. Indeed, capabilities are created through a complex combination of *paths*, *processes* and *positions*.[25] In simple words, managerial creativity and enterprise shape the creation of these capabilities as they leverage existing or new resources, to structure creative and innovative routines. Some of it is historical and path-specific, making the way capabilities are built unique to each organization.

In summary, when defining an instrumental purpose, rethink an organization as a bundle of digital capabilities. Any organization's ability to leverage technology depends on its ability to visualize and conceptualize organizational digital capabilities that will help it realize identified goals. I outline the ways to visualize and conceptualize digital capabilities in the next chapter.

4

Visualizing and Conceptualizing Digital Capabilities

'Going ahead, in the era of intelligent cloud and intelligent edge, you are going to see a ubiquitous computing fabric that is distributed. You are going to have intelligence that is ambient, experiences that span devices and senses. To me, it comes down to how every organization here in India can ride this wave and build their own tech intensity.'[1]

—Satya Nadella, CEO, Microsoft

'The most important consideration is, you have to build your own tech capability.'[2]

—Satya Nadella, CEO, Microsoft

While defining the instrumental purpose of digital transformation, managers must identify the digital capabilities to be built. Building high-performance digital capabilities is a complex process, primarily because visualizing and conceptualizing these capabilities is challenging. Even for established leaders, conceptualizing high-performance digital capabilities is not easy. This is because many times these capabilities are revealed unknowingly. Digital capabilities built in one context may provide the visualizations and creativity required to build such capabilities in another context. Sunder Pichai, CEO of Google and Alphabet Inc., narrates Google's experience with the same:

> Google's efforts in India have deepened our understanding of how technology can be helpful to all different types of people. Building products for India first has helped us build better products for users everywhere . . .
>
> A recent example is GPay, our fast, simple way to pay contactless or online. Together with the rise in BHIM-UPI adoption, GPay makes it easy to pay the rickshawala or send money to family back home. India is setting the global standard on how to digitize payments, and it's now helping us build a global product.[3]

Visualizing and Conceptualizing Digital Capabilities

Why is it important for managers to learn how to visualize and conceptualize digital capabilities? Primarily to decide what to build, if at all. Managers are often faced with questions such as: Do I need to create a digital capability (say, for communication) when my employees are already good at working (communicating) without technology? Well, the underlying assumption while defining an instrumental purpose is that the digital capabilities being created are far superior to traditional capabilities. For example, ChatGPT may help create capabilities that are alternatives to the human interface for customer service. The decision criteria then is whether the AI-based large-scale language models (such as ChatGPT) are better performing than the traditional capabilities for customer services. So, while choosing digital capabilities to build, managers assess whether these capabilities have far greater impacts on performance. However, ascertaining the performance impacts of these digital capabilities requires that one visualizes their performance impacts. This visualization has not been easy.

Digital technologies existed long before organizations realized their effects on performance. However, the visualization of their impacts has not been easy. In fact, for a long time, many argued against their potential competitive effects. In his 2001 *Harvard Business Review* article titled 'IT doesn't matter',[4] Nick Carr argued that there is no need for strategic emphasis on IT. In his comparison, digital technologies are similar to electricity. So, he concluded that people managing these should be relegated to the basements and not the C-suite. In my opinion, this lack of understanding for conceptualizing and visualizing digital capabilities was not apparent at that time. Subsequently, for a long time, I argued against this point of view through my research. With my co-authors, I showed that realizing the instrumental purpose of digital transformation requires a vision—a deep organizational one. Indeed, contemporary organizations have demonstrated tremendous visualization in creating high-performance digital capabilities. Notably, unlike electricity, successful organizations have championed digital transformation to focus on apt performance goals.

The Capability Performance Hypotheses: The Performance Visualization

What are the performance goals for aspiring to build digital capabilities? Broadly, creating high-performance digital capabilities focuses on two goals: innovation enhancement and cost reduction. At any time, the organization has to balance the two. I argue that one has to use the evolutionary perspective to leverage the instrumental purpose.[5] First, the view underlines the role of innovation. Innovation is crucial as the world—consumers, stakeholders, suppliers, technology, competition and employees—is constantly evolving. The development of a (digital) capability underlines that organizing (with technologies) is an evolutionary phenomenon. That is, innovative, high-performing digital capabilities further evolve. Evolution is a process that has manifested over the years, and it indicates moving down a path that links the early-age chemical soup (the form of the earth) to the present stage of individuals.[6] Organizational (digital) capabilities behave similarly, as they help the organization evolve in a competitive environment. Indeed, the notion of competition is inherent to the evolution process. Most organizations face competition for resources. However, at the organizational level, evolution is a collaborative process as well. Lions may hunt buffalo in packs. Similarly, organizations may compete (for resources) and collaborate (to harness synergies) simultaneously. Innovative digital capabilities offer a competitive edge and may help organizations collaborate.

Second, the goal of digital capabilities may be to reduce costs. While the organization as a mechanism for innovation is indeed part of the story, it is *not* the entire story of digital transformation or evolution. The organization is a mechanism for reducing costs. Evolution manifests as we consume intelligently, thereby reducing the costs (say, of our shopping basket). Consider how customers reduce their shopping costs. A consumer may use many ways to seek recommendations for products. Traditionally, people may talk among their friends' circles or seek recommendations from acquaintances. However, getting these recommendations is time-consuming, and getting perfect recommendations (based on friends' or acquaintances' use of multiple competing products) is nearly impossible. The travel and search costs for such a project are overwhelming. Digital technologies have changed the paradigm for seeking recommendations. For example, Amazon's digital capabilities provide recommendations to millions of customers for each purchase. These reduce the costs for consumers compared to the effort involved in seeking face-to-face recommendations.

Similarly, greater performance has manifested for organizations that have developed capabilities to automate shipping and delivery, create innovations in delivery (such as real-time location or tracking) and so on. Some organizations use these to purposefully transform (reduce) their costs of operations. Many digital transformations are helping reduce costs, as crucial time spent tracking or providing customer service is reduced. Previous human capabilities are now being replaced by digital capabilities, built through advanced servers, databases and mobile and web apps. That is, the world has moved on from the days of Carr's predictions. Why would building certain capabilities be the key to enhanced performance? This is not an easy question to answer. It requires managers and students to test the capability-performance hypotheses. Over the last few years of my research and executive interactions, I have found the need for managers to test a *capability performance hypothesis* in order to realize the instrumental purpose of digital transformation. It helps assess the performance impacts of digital capabilities.

While greater innovation and lower costs are important criteria, organizations still have to choose from competing digital capabilities. The capability performance hypotheses require visualizing: which digital capabilities should the organization champion and build, and which should it not?

Traditional vs Digital Capabilities

A comparison of digital capabilities with traditional capabilities offers insights into how to visualize high-performing capabilities. The first realization any manager has to make is that traditional capabilities (that are being considered for digitization) are already driving performance. However, in the longer term, digital capabilities, when their potential is realized, have a greater performance impact (see Figure 4). To see the long-term potential, think about the transformation of food organizations over the ages. Imagine life a really long time ago, say in the early Paleolithic Age. Back then, humans had to venture out and source food, and this food had to be consumed instantly (as there were no refrigerators). For most of our existence, humans were unable to store food in the supply chain. Now think about life today. Things available to eat are abundant. Enhanced organizational capabilities enable the management of food throughout the value chain, starting at the farm and ending at dinner tables. Today, large organizations are building advanced digital capabilities for sustainable farming, nutrition, security and safety in food production and consumption. For example, in the US, Walmart is leveraging blockchain technologies to transform the way it collects and processes information. This helps enhance security in the food supply chain. In the spring of 2018, many people got sick after consuming contaminated romaine lettuce due to an outbreak of E. coli bacteria. Following a two-year pilot after the incident,

Walmart used blockchain technology to be able to track every bag of spinach and head of lettuce. So, a recall could be initiated for specific suppliers, instead of recalling all supplies. Newer technologies create the potential for many such advanced organizational abilities.

ChatGPT is one such technology. In 2022, Bill Gates challenged the OpenAI team, which was building ChatGPT. He asked the team to prepare the software to pass an advanced placement (AP) biology exam without training it to answer specific biology questions. ChatGPT soon received a near-perfect score and missed only one mark out of fifty.[7] In other words, ChatGPT stands out in comparison to the human mind. This comparison with the traditional capabilities is the first step in visualizing digital capabilities.

To do this comparison, a simple assessment of digital capability potential is often the key. This digital capability potential is calculated by comparing it with traditional capabilities (for doing the same thing). Specifically, managers need to assess the following:

[Pr (Dig Cap)–Pr (Traditional Cap)]/ Pr (Traditional Cap)

where,

Pr (Dig Cap): Performance of digital capability
Pr (Traditional Cap): Performance of corresponding traditional capability

That is, conceptualization and visualization entail assessing the performance of digital capability compared to that of traditional capability.

Temporal Dynamics: Potential vs Realized Digital Capabilities

Comparing digital and traditional capabilities is not straightforward. Digital capabilities are not high-performing from day one, while traditional capabilities are usually equally effective across time. Indeed, any investments over the lifetime requires taking into account the timing, to penalize delays in returns. Often, net present value (NPV) or a similar model is used to discount the delayed returns. A contrasting dynamic prevails in the case of digital capabilities, as their returns increase over time. In comparison to traditional capabilities, potential of digital capabilities is latent, as they enhance performance only over time. An example of this is digital customer reviews. Many businesses (such as those selling books, groceries, movies and so on) enable customers to leave reviews, based on their experience. The digital capabilities to seek reviews require technology for customers to add their assessments, rate them, leave a video or image and similarly share other information. However, this digital capability is not instantaneously rewarding. Customers find the capability valuable only over time, after enough reviews have been added for a product. Similarly, the development and training of a recommendation algorithm is costly (though it may have near-zero replication costs) as its rewards are realized only when enough data is available on users buying behaviours, preferences and characteristics. So, to visualize the performance capability, one has to separate the potential at any point in time and may use an apt discounting method to assess the capability performance vis-à-vis a traditional capability.

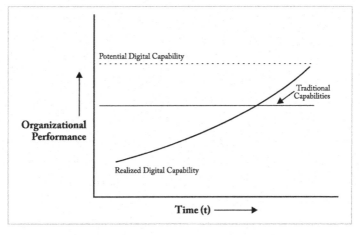

Figure 4: Potential vs Realized Capabilities

The discussion so far underlines that visualizing and conceptualizing is standardized. That is not the case. Visualizing high-performing capabilities is an organization-specific phenomenon, which means the same technology may be used differently. Further, any two organizations may realize different returns from the same capabilities. Visualizing and conceptualizing the performance effects requires assessing the following two dynamics:

1. How the organization leverages technologies by building digital capabilities, as assessed through the Digital Capability Conversion Factor (DcCF)
2. The relative importance of digital capability in the organizational context, assessed through the Organizational Capability Performance Factor (OcPF)

The two—DcCF and OcPF—manifest as a result of the fact that capabilities are organized and created in a hierarchy and they have different performance impacts.

Hierarchical Capabilities Strategies

Most digital capabilities are a driver for bigger and broader organizational capabilities. Digital technologies (or other digital capabilities) may be the lower-order for higher-order organizational capabilities. Researchers often define this notion as the hierarchy of capabilities, in which some lower-order capabilities lead to higher-order capabilities. A simple example illustrates this phenomenon. As a child, one's ability to learn the alphabet helps them create higher-order combinations—words, sentences and then paragraphs. In turn, these may lead to even higher-order capabilities for writing Booker Prize-winning books or Nobel Prize-winning research manuscripts. Similarly, digital capabilities enhance performance when they lead to the creation of high-performing organizational capabilities.

Consider the case of recommendation agents who can suggest what customers may buy next, by analysing data on millions of customers' purchase of multiple products. Such a digital capability to recommend enhances a retailer's customer service capability, which in turn enhances performance (say, by enhancing customer loyalty or repeat purchases). Therefore, the question of which capabilities to build relies on the ability of the organization to think about and champion the hierarchy of capabilities. Deep research forms the foundation for this dynamic.

I tested the hierarchy of capabilities in the customer service domain across branches of a large Indian bank. Specifically, along with my co-authors, I studied how a branch's operations can enhance its customer satisfaction. This research was published in *MIS Quarterly*—a leading journal—around ten years ago,[8] and it continues to attract attention from researchers and practitioners across this domain. The research analysed whether the digital design—the quality of the information in the branch—may change the abilities of branch employees to discern customer needs and fulfil them. We studied four aspects of information quality: how accurate the information was, how complete the information was, how current the information was and in what format the information was available to the branches. More importantly, simply providing this information to the employees was insufficient.[9] Our findings revealed that digital design was a lower-order capability that led to higher-order customer service capabilities.

The improved digital design specifically enhanced the branch's customer orientation and customer response capabilities. Customer orientation means that branch employees are curious and oriented towards customer needs. In such branches, employees continuously think about customers. There is a culture that is customer-oriented. Customers are the centre of strategy and operations in customer-oriented branches. Second, employees in these branches demonstrated greater customer responsiveness due to their improved ability to serve their customers. The two capabilities did not originate on their own and were the result of a development process following the hierarchy of capabilities. Furthermore, the effects of digital design differed across the branches. The ability to convert lower-order digital design into higher-order customer service capabilities (for customer orientation and customer response) varied based on their process sophistication. Any bank may have to prioritize where and in which branches they would like to build these capabilities and in which branches they would not like to build them. In summary, the hierarchy of capabilities helps the organization visualize and conceptualize the performance effects of digital capabilities, and DcCF and OcPF help assess these effects.

Mastering the Instrumental Purpose: Digital Capability Conversion Factor (DcCF)

First, mastering the instrumental purpose requires that managers visualize or assess the DcCF. The digital capability conversion factor represents the extent to which a digital capability contributes to a specific organizational capability. It is assessed as:

$$\triangle(OrgCap)/\triangle(DigCap)$$

It is intuitive to think that each digital technology (or capability) may have a different effect on an organizational capability (see Figure 5). For example, if we consider technologies with material requirement planning (MRP) and customer relationship management (CRM) functions to represent two different lower-order digital capabilities, their effects would vary across higher-order digital capabilities they build. Consider their impacts on manufacturing responsiveness and customer service capabilities. The MRP capabilities of an organization may enhance the manufacturing responsiveness capability much more than they may influence the customer service capabilities. Thus, the DcCF for MRP capabilities is very high for manufacturing responsiveness but low for customer service capabilities. Alternatively, DcCF for CRM capabilities is high for customer service capabilities. So, managers must visualize different capabilities in terms of DcCF. When evaluating the capabilities performance hypotheses, plot the DcCF over time for different organizational capabilities and digital technologies. The manager can also account for any costs incurred due to performance delays,. Indeed, the costs may include costs of investments in technologies, or social and organizational costs, and most importantly, opportunity costs.

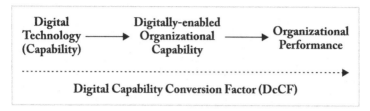

Figure 5: Hierarchy of Capabilities

Organizational Capability Performance Factor (OcPF): Role of Context

To assess the role of context while evaluating DcCF, the organization has to evaluate the OcPF. The OcPF assesses how valuable a certain capability is for a specific organization. Formally, it assesses the marginal effects of an organizational capability. That is, how much of an increase in performance is due to a unit increase in the organizational capability (see Figure 6). Often, the differences in the context of a firm may lead to differences in organizational performance. Previously, researchers have indeed identified capabilities according to the context. Context has been found to play a significant role in shaping the impact of knowledge absorptive capacity and customer orientation and response capabilities. My research has significantly contributed to the assessment of these context-dependent impacts of digital capabilities. For example, the customer orientation capability (discussed above) may have different performance impacts for different branches based on their process sophistication. In general, different capabilities matter differently for organizations. For example, a firm that has its performance driven by operational excellence (e.g., Foxconn, the electronics contract manufacturing company and the biggest supplier to Apple) will realize greater performance due to digital manufacturing capabilities rather than digitally-enabled customer orientation capabilities. Because it acts as a backend provider serving only a select few customers, Foxconn may not gain much from its customer orientation capability. Therefore, its OcPF is low for the customer orientation capability. In comparison, Amazon has a very high OcPF for customer orientation capability, as it helps

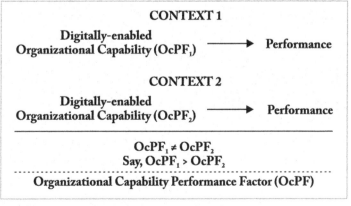

Figure 6: Organizational Capability Performance Factor (OcPF)

serve a lot of customers better (and that is the key to high performance for the retailer).

Conclusion

In life, our goals are designed by outsiders (society, parents, school, organizations), who may differ for each individual. The fulfilment of the goal entails a focus on performance. The instrumental purpose—to build high-performance capabilities—requires a specific focus on performance outcomes: innovations, cost savings or both. The focus on instrumental purpose underlines that one may *not* equate digital transformation with adoption of advanced technologies. A view on adoption of a specific technology—such as e-commerce technologies, blockchain technologies, AI, machine learning or analytic technologies—makes it myopically technology focused. Following the technology view, at best, an organization acquires standardized technology implementations. This vastly undermines the organizational ability to assess the potential business value of digitization for them. Many technologies have features that overlap with existing work methods or current technologies in use. Instead of focusing on specific technologies, digital transformation requires a focus on the organizational use and value of digital technology. This organizational view, instead of a technology-focused view, is transformational. High performance manifests when one transforms how one is organized. This requires visualizing and championing appropriate digital capabilities. The chapter outlines a hierarchy capability process—defining a DcCF and an OcPF—that organizations may use to visualize and conceptualize high-performance digital capabilities.

Part 3
The Operational Purpose

5

Operationalizing Digital Transformation

How do we digitally transform? Answering the question is not easy. Seasoned managers, scientists, strategists and planners have found it hard to transform digitally. Reports indicate that 70 per cent of digital transformations fail.[1] Costs and return on investments (ROI) concern anyone undertaking a digital transformation. These concerns usually manifest when an organization does not have a clear model for its digital transformation. This model offers a strategy and vision to transform digitally and is required to transform successfully. Even large organizations with long experience in transformation tend to fail. This is largely because digital transformation requires something unique, over and above managerial skills. It requires us to rethink how we organize. While digital technologies transform human organization, undertaking this transformation has often been challenging for organizations. Ford is one such example.

Ford initiated a smart mobility enterprise to counter the growing environmental concerns associated with automobiles. The chairman, Bill Ford, underlined worries regarding the increasing parc of vehicles reducing his motivation to sell more cars.[2] Ford's smart mobility is centred on saving the environment and creating autonomous vehicles, in addition to many other smart mobility solutions. That is, Ford planned to leverage modern digital technologies, say for data and analytics, to offer advanced mobility services.[3] Ford knew *what* it wanted to do. For example, it wanted drivers to be able to remotely access their vehicle features so they could remotely start the car—a great capability in freezing conditions—and unlock its doors or see the parked vehicle's location on a smartphone app. Similarly, it wanted to build FordPass® as a free digital, physical and personal platform. This platform would offer members a marketplace for mobility services, enabling better parking and sharing, among other features. Similarly, *what* it wanted to do with data and analytics was exemplary. Ford collaborated with IBM to offer research scientists small data (ten or fifteen seconds long), which they could use to find patterns and trends for building smart applications that support Ford's experimental Dynamic Shuttle—a vehicle to offer on-demand shuttle service for its employees in Dearborn,

Michigan.[4] The offering was part of Ford's smart mobility digital initiative. The broader initiative was not simple to implement. While the 'what to do' was clear, the 'how to do it' remained elusive. So, Ford faced challenges with its digital transformation through smart mobility, resulting in frustration.[5] As a result, the stock price plummeted soon after and the CEO had to resign.

This is not an isolated case. Digital transformation is challenging for many organizations. Often, legacy systems and processes may slow down an organization's performance. More recently, cyber security threats may hinder the ability of the organization to operate normally. Further, employees may often resist digital transformation, especially when it involves numerous changes in the work processes. SAP implementation is a very telling example. SAP is known for its enterprise resource planning (ERP) systems and implementing them requires vast changes in organizational work processes. Many companies have faced challenges with these implementations. Hershey's, the American multinational brand largely famous for chocolates, cookies and beverages, is one such example. The company planned a digital transformation through an ERP system to replace its legacy systems. The infamous Y2K problem triggered the change. Some of you may remember the serious concerns before the year 2000 that the turn of the century and millennium could hit legacy systems hard. To put it simply, the problem was with the way the year was written in code. Instead of four digits (say, 1998), many lines of code used two digits for the year (say,

98). Many problems were projected to happen because of this, though eventually much less happened.

Hershey's management wanted to move the date of its ERP implementation delivery earlier to avoid the Y2K problem. They had thought it would cost about $112 million in approximately four years. However, the decision to move up did not go very well. Hershey's had to suffer, and what made it worse was that it all happened during Hershey's busiest business retail season. What they wanted to do was clear, but they weren't sure how to go about it. As a result, Hershey failed to ship products to the retail stores. It was nearly fatal for Hershey.[6]

Indeed, while many organizations see opportunities in building digital capabilities, they are often hindered by challenges while transforming digitally. And this is not the case with private organizations alone. Digital transformation has also been challenging for public sector organizations, such as the US government's Affordable Care Act (ACA) website launched in 2013,[7] when former president Barack Obama was at the helm. Signed into law around 2010, the ACA was designed to extend health coverage to millions of Americans who didn't otherwise have insurance. The idea was to build a digital health insurance marketplace, so no insurance company would be denying coverage to anybody due to a pre-existing condition. Many a time, healthcare is largely dependent on insurance availability. The insurance companies can take into account whether you have a pre-existing condition (and other such characteristics), shutting many people out of the healthcare sector. The website was crucial to operationalizing the act, but it failed despite many efforts. Several issues hindered its launch, causing visitors to the website to experience long waiting times, encounter error messages or receive incorrect information. Things deteriorated so much that Obama had to address the issues and promise to fix them quickly. Financial constraints were not a reason for these debacles. The budget for the website was approximately $850 million.[8] However, the news is not all bad. Not everyone fails. There are examples of successful digital transformations. One of them is DBS Bank's digital transformation.

The Case of DBS Bank

To understand a successful digital transformation, think about the transformation of DBS Bank, especially starting in 2014. At that time, DBS Bank was one of the leading banks in Asia. It was the only Asian private bank on the list of the top ten banks in Asia, with the list largely dominated by western banks. DBS is a Singaporean bank with operations in Greater China, South Asia and Southeast Asia. In 2014, the bank had 280 branches in eighteen countries, offering a broad spectrum of services in institutional banking, wealth management and consumer banking. It also had an extensive network of about 2500 touch points, comprising bank branches, ATMs, self-service kiosks, etc. At that time, it was one of the strongest and most well-capitalized banks in the world and was operationally renowned. Many accolades underline this operational excellence. Till 2014, it had already been ranked the sixth safest bank in Asia for six consecutive years.

It was doing well when it realized the need for digital transformation. DBS Bank was facing growth constraints due to the advent of new nimble competitors and a failed takeover of Bank Danamon Indonesian in 2013. The changing demographics of customers also made it necessary to transform digitally. The new generation of customers was much younger and very tech-savvy. So, DBS Bank decided to digitally transform to become a new version of itself. This was necessary in order to survive and thrive in the next era. The digital transformation that unfolded gives a glimpse into many ways in which an organization may leverage technology. Many unique and remarkable elements of digital transformation emerged, and I will elaborate on these in a more formal model (the DaWoGoMo© model) that underlines the operationalization of digital transformation in this part of the book.

There were many challenges for DBS Bank's transformation. It was a large bank with many touchpoints and a well-established set of processes and operations. More importantly, it had a workforce that was not digitally savvy. So Piyush Gupta, an IIMA alumnus and head of DBS at the time, thought of a digital transformation strategy. For this reason, the bank championed several initiatives. To streamline the process of digitization across divisions, DBS merged two of its core divisions: technology and operations. Not only did it lead to the rapid digitization of processes, but it also enhanced the alignment between technology and operations. The new digital work processes were more in tune with the new business environment. Many other noteworthy transformations characterized the bank's digital transformation.[9]

1. The bank put the customer centre stage, asking questions such as: What are the customers' needs? What are the innovations that should be championed to enhance the customer experience? How do we create innovation?
2. Many start-ups were looked at to boost internal energy and motivation for digital transformation. Boot camps and hackathons were held to discover and unravel the steps that the company would like to take.
3. DBS launched different mobile apps, including the DBS PayLah, to book tickets and rides, order meals, manage card reward points and so on.[10] Similarly, a Home Connect app enabled young customers to find out the price of a particular property through the app itself, as well as assess EMI, other costs or a mortgage.
4. Internally, DBS revamped its technology through the IDEAL platform that enabled Internet banking.[11]
5. DBS harnessed data, as it created an analytics competency centre. It came up with the idea of using voice analytics to analyse various calls. This helped them identify what customers were complaining about and what it was that they were happy with. This helped them increase compliments and reduce complaints.

6. To spread the word about this new way of banking, DBS Bank went through an extensive social media campaign in India. Classified as 'Chilli Paneer',[12] DBS made several films around food, cricket, Bollywood and other topics relatable to a vast set of Indian banking consumers. Similarly, localized social media campaigns were launched in other countries.
7. It worked hard to change the mindset of its employees, transforming them from 'Technology Idiots' to 'Digital Warriors'. Many initiatives helped achieve this. One of them was 'Tell Piyush', whereby employees gave suggestions, asked questions or just commented to Piyush, getting engaged with the transformation.

Many of these are crucial elements that outline the successful operationalization of a digital transformation.

Operationalizing a Digital Transformation: The DaWoGoMo© Model

How do organizations craft an operational plan to transform digitally? The digital transformation challenges the current state of an organization's being. Creating a new 'digital' organization requires a clear operational plan and purpose, outlining how an organization will be transformed. I offer the DaWoGoMo© model that outlines four aspects for the successful digital transformation of an organization. Any organization would need to master the transformation of these four elements (and a fifth element in an extended DaWoGoMo© model, which I will mention later).

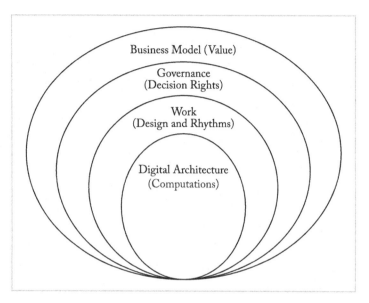

Figure 7: The DaWoGoMo© Model

The first aspect of the DaWoGoMo© model (see Figure 7) is digital architecture. That is, managers have to transform digital technologies and their relationships. The second element that organizations have to transform is work. We would fail operationally if we forgot that the core aspect of an organization is work. What is work and how may it be transformed? Digital transformation requires that organizations think about these questions deeply. The third element of the DaWoGoMo© model, the 'Go' part of the model, is governance transformation. Any organization is foremost a governance mechanism. Even in homes, parents have to govern their children. They must set rules and define methods and procedures. The governance of other organizations is a bit more complex. How should managers govern the organization differently? What aspects require different governance? These are some of the questions managers must answer as they transform the third element. The last element of the DaWoGoMo© model entails transforming the business model. Every organization has a business model. There is a certain set of activities it performs, and there are certain values attached to those activities. An organization transforms as it continually tries to maximize its value. Digital technologies transform the value of activities. Who does the organization interact with and how? What is the value that emerges from these interactions? These questions have to be thought about again by the organization. While business model transformation is the last element of transformation, we may only be successful in the operational purpose if we clearly understand the people aspect of the organization. To assess the influence of people and their motivations, I underline a cultural transformation, in addition to the DaWoGoMo© transformation. This extended DaWoGoMo© model defines the basis for the operational purpose that underlines any organization's digital transformation.

6

Transforming Digital Architecture

'Every half a second, a child goes online for the first time. Children and young people are engaging with the digital world in unprecedented breadth and depth. We need to catch up—and invest in digital infrastructure, solutions, and standards so that they can safely and positively engage with the digital world. We also need to bridge the digital divide and reach the 2.9 billion people who remain offline with equitable access to online tools and services.'

—Catherine Russell,
executive director, UNICEF[1]

Transforming Digital Architecture

Digital architecture is emerging as a radical influence on our lives. For example, in India, a UPI-based architecture for payments has transformed how people transact and make payments. Across businesses in India, these digital architectures have transformed lives, as customers prefer to use various digital payment apps to transact instantly. UPI-based architectures enable these transactions. Traditionally, payment architectures were heavily driven by credit card-based systems that often had hefty fees. Small traders may avoid using these systems or have to bear the transaction costs. The use of nimble and agile digital public infrastructures, such as UPI, is transforming lives and well-being, across the world's population. One of the leaders of the digital revolution, Bill Gates, underlines the tremendous social and global impacts of the modern infrastructures developed by public organizations: 'There is not a single Sustainable Development Goal that digital public infrastructure won't advance in one way or another. It is amazing in international development when one targeted investment can have spillover effects in almost every issue area we care about.'[2]

Bill Gates is not the only one to underline the role of digital architectures. David Malpass, the American economic analyst who served as president of the World Bank Group (WBG) says:

> There have been severe reversals in development, and global policy trends suggest these will persist. Digital public infrastructure is a vital part of our response. Digital identification, payment, and data sharing platforms have made it possible for countries to respond

more effectively, more transparently, at a greater speed and scale, and with more security and privacy. For the digital transformation to be successful, we need trusted, quality, and inclusive public infrastructure, accessible and affordable Internet, and the development of digital skills.

While there is a lot of enthusiasm and curiosity today about digital public infrastructure (DPI), it is notable that digital architectures have been most prevalent within private organizations. In older organizations, large-scale systems set the standards for high performance. Notably, the American Hospital Supply Corporation's (AHSC) Analytic Systems Automatic Purchasing (ASAP) system and American Airlines' Semi-Automatic Business Research Environment (SABRE) system are renowned digital architectures. First, ASAP was renowned for its ability to automate many tasks associated with purchasing processes. The system leveraged extensive data processing. It helped maintain historical and real-time inventory data, as it facilitated automated tracking of orders, inventory levels and shipping information. Furthermore, the system was highly customizable, making it easy for employees to adapt it to specific user needs.[3] Second, the SABRE system by American Airlines enabled real-time access to flight information, reservations and data for airline personnel. The architecture transformed the organization of various airline functions: flight scheduling, reservation management and pricing.

Even today, digital architectures form the backbone of most organizations. These architectures may include ERP systems or CRM systems, among others. ERP systems help digitize and manage business processes and CRM systems help the organization manage customer relationships and create personalized services. Through CRM systems, organizations manage and store large quantities of customer data. Both of these systems provide a unified view, advanced analytics, reminders and other functionalities to enable employees' decisions and actions. However, unlike in the past, the landscape of digital architecture is very complex. Today, it is a bit challenging to think about what digital architecture means.

What is Digital Architecture?

Digital architecture is a complex construct that has been challenging to conceptualize and define. Various scholars have conceptualized digital architectures differently. Peter Weill and Sinan Aral define four types of IT systems, outlining the composition of digital architectures: transactional, informational, strategic and infrastructural.[4] Transactional systems facilitate partner and customer transactions, while informational systems provide comprehensive insights to management, such as accounting, customers or compliance-related information required for reporting. On the other hand, strategic systems support planning and innovation, empowering organizations to reflect on their strategies and develop new products and services. Lastly, infrastructural systems serve as the backbone for other digital systems. These ensure seamless connectivity between computers. And, as it is colloquially called, they are crucial to keeping the lights ON. Some scholars have identified two distinct aspects of digital infrastructure: technological infrastructure and human IT infrastructure. My previously published research has shown that these two aspects of digital infrastructure perform differently. Weill and Aral (2006) underline a case (Considering IT Investments as a Portfolio, by CISR at MIT) that an average company spent approximately 46 per cent on infrastructure, 26 per cent on transactional systems, and 17 per cent and 11 per cent on informational and strategic systems in 2005.[5]

Often, organizations need to analyse their business environment to create an optimal digital architecture. The two components that make up digital infrastructure highlight its dynamic nature. Studying data from 355 unique firms in the USA over three years, we found variations in digital infrastructure across business units of large firms. Further, my co-author and former doctoral student, Franck Soh, and I discovered the effects of a dominant digital infrastructure configuration across the business units of a firm.[6] Our findings indicated that organizations may have to transform (or monitor) the changes in their digital infrastructure. Specifically, organizations have to create a more human-led (or balanced) infrastructure when business conditions change rapidly. Why do these effects manifest? Because digital architectures and organizations' performance are intricately intertwined. To understand the deep relationship between the two, it is imperative to see the organization from an information-processing perspective.

Organization: A Computational System

Digital architecture forms the computational foundations for the organization, which itself is a computational entity. Seeing an organization as a computational system is a well-established paradigm in organizational theory. In the late twentieth century, Galbraith underlined that an organization is a structure for information processing.[7] According to this perspective, an organization processes information internal to the organization or from the external environment. Since environment changes are constant and threatening, its ability to do so is crucial to its survival and growth. Metaphorically, an organization is like a fish in the pond, surrounded by flora and fauna, some of which are hostile and others are the source of energy. The fish has to compute to be able to discern what to consume and what to avoid. Similar to the fish in a pond, the organization lives in a broader environment from which it consumes, to which it belongs, and which gains from the organization's presence as well.[8] However, unlike the fish in the pond metaphor, an organization and its environment are much more complex, comprising customers, regulators, competitors, potential employees, technologies, universities and so on. To thrive in the environment, an organization has to collect, process and disseminate a lot of information. Digital architecture forms the computational backbone of an organization, enabling its information processing. I will elaborate more using the example of a hypothetical flour mill.

Consider a hypothetical flour mill that faces competition in its neighbourhood. Seeing it do well, many others have opened mills in the vicinity. Which of the flour mills will eventually attract more customers and perform the least-cost operations to become more profitable? If we think about the organization as a computational system, the winning flour mill may try to collect maximum information about the customers, changing regulations, new trends in technology and so on. It then develops an efficient system to analyse this data, thereby transforming its internal information processing methods. For example, an online ordering system may be created if customers expect to order using their mobile devices. Such digitally-enabled exploitations offer an edge over their competitors, as they reduce the costs for the customer to travel and spend time ordering at the mill.

Digital architectures are also catalysing the emergence of modern digital organizations. Consider the case of modern hospitality companies, such as Airbnb or Oyo. These online platforms offer customers easy access to lodging (hotel or housing) across India and the world. Consider the information processing involved in these hospitality companies. Booking accommodations is a complex task that involves real-time information processing and inventory management. A room cannot be booked by two patrons at the same time, so real-time updates are required for a smooth booking experience. Further, the information involved in the business is complex and includes real-time pricing, deals, lodging characteristics, customer or demand-related data and so on. In addition, various algorithms process this information for pricing adjustments (say on holidays and other more popular travel dates), forecasting demand, optimizing inventory, customer delight through quality offerings and many other related business processes. Modern technologies, such as those leveraging machine learning, are crucial for these organizations to create a seamless experience for their customers.

Digital architecture's impact on organizational performance is now seen all across the board. Consider retail giant Amazon's Go stores. Amazon Go is a fully automated retail store with no cashiers. AI with cameras track each movement of the customer as they enter, scanning their face and linking it with their Amazon account. Subsequently, all their behaviours are logged in and an automatic checkout happens through the app. Compare this with the local brick-and-mortar stores, such as your neighbourhood grocery shop. Both of these entities are involved in information processing, but their approaches differ significantly. Amazon has a substantial advantage due to its very advanced digital architecture, which enables it to process a large quantum of information. This extends far beyond what is feasible for a physical store, particularly regarding customers' shopping behaviour. While a local store might employ cameras and other methods to monitor customer activity, the level of detail Amazon obtains is exponentially greater. Through its Amazon Go stores, Amazon is privy to information regarding the number of items customers place in their shopping carts and those they subsequently remove. Similarly, through online retail stores, Amazon may also know the items customers add to their wish lists. An advanced digital architecture may enable Amazon to perform in-depth analyses, allowing it to craft various promotional campaigns, personalized advertisements and enticing discounts tailored to customer preferences and shopping habits. That is, digital architecture offers Amazon an edge as it enables the retailer to provide a highly personalized and data-driven shopping experience. It is nearly impossible for physical stores to match the experience. This is a testament to the transformative potential of advanced information processing in the retail sector.

This information processing spans public organizations as well. The information processing perspective is at the core of many extraordinary human achievements. Consider the case of an organization that you most likely know about, the Indian Space Research Organization (ISRO), which is renowned for its remarkable accomplishments in launching satellites, the most recent being Chandrayaan-3 with the Vikram lander on the south side of the moon. Satellite launching, too, requires large amounts of real-time information processing. During launch, the information processing may entail evaluating the satellite payload and launch vehicle speed and trajectory. Such information processing involves many different sensors and other hardware that collect real-time data during the launch. Information processing in real-time is the key to enabling decisions leading to a successful launch. Both the launch vehicle and the satellite are in constant communication with ISRO, as the organization receives and processes data transmitted by them. In-orbit satellites engage in continuous real-time information processing with the organization as well. Some of this data is invaluable for developing innovative satellite technologies, as exemplified by the Mars Orbiter Mission, which facilitated the study of the Martian atmosphere, or Earth observation technologies, including synthetic aperture radar (SAR) and hyperspectral imaging.

Organizational information processing pervasively spans the day-to-day lives of individuals, too. Beyond large organizations like Amazon, Airbnb or Oyo that use bigger systems, such as ERP and CRM, individuals are transforming how they process information. They buy mobile phones and install UPI apps for payments, use WhatsApp for communication with family, order groceries online and use apps for relationships, schooling and so on. Large organizations are also transforming individual lives through their digital architecture. For example, ISRO leverages its Earth Observation satellites to serve the public. The satellite collects data from the Earth's atmosphere. Then, technologies combine information from ground-based sensors, weather models and other sources. This information processing enables individual navigation, weather forecasting and disaster management. Historically, information processing has helped humans evolve. Early on, humans had to venture out in search of food. Inventory-enabling technologies have helped humans evolve a new way of life. Today, one may walk into a retail store, go to a roadside vendor or go to a restaurant and get access to food instantly. Modern-day retailers and others use technology to manage inventory throughout the value chain, starting at the farm and ending at dinner tables. As I will argue in Part 4, information processing spans even the smallest of organizations and underlines the existential linkages between life, organization and digital transformation.

Digital Architecture and Organization

Digital architecture transformation is required to capture the new possibilities for information processing that have become plausible with the advent of technologies such as AI. Not surprisingly, a successful digital transformation has its heart in the transformation of digital architectures (the Da in DaWoGoMo©). However, shouldn't transforming digital architecture be straightforward? Why not just adopt all digital technologies as they become functional? Managers are faced with a unique problem while carrying out digital architecture transformation—the problem of 'plenty'. They must continuously decide which new technologies to adopt. The advent of many advanced technologies, like the Generative Pre-trained Transformer (GPT-3) is an example. Advanced neural network models have given rise to technologies such as large-scale language models (e.g., ChatGPT). At large, many such advances in the domain of AI, notably, are threatening the growth and survival of organizations. However, they are also opening new possibilities. Is 'plenty' really a problem? Can the organization not just ignore the fact that there are too many new digital technologies? Why shouldn't they ignore it? While organizations are mindful that the advent of innovations compels them to ponder how to process information differently, digital architectures are complex combinations of various digital technologies. Changing the information processing model is not easy, as it may not be the best way to enhance the information processing capabilities of the organization.[9] That is, merely adding or deleting new technologies may deteriorate one's ability to process information. And an organization must think *wisely* about the transformation of digital architecture. Fortunately, much of this wisdom exists, albeit in traditional architecture theory.

The 'Traditional' Architecture Wisdom

By definition, digital architecture is 'architecture', as it supports organizations. Organizations have been manifested around architecture for ages. Consider temples in south India. These have concrete architecture and a set of social organizations happen around it. The two are intricately intertwined. This was not incidental, and a pattern is relevant because it was designed with the wisdom learnt over the years. A core architecture embeds wisdom. Therefore, the great architecture that embodies this wisdom made the organization perform better and stand out among peers. Such wisdom goes beyond specific styles or forms. L. Corbusier (2017) [10]summarizes this when he says:

> Architecture has nothing to do with the various 'styles'. Louis XIV, XV, XVI and Gothic are to architecture what feathers are to a woman's head; they are pretty sometimes, but not always and nothing more. Architecture has graver ends; capable of sublimity, it touches the most brutal instincts through its objectivity; it appeals to the highest of the faculties, through its very abstraction. Architecture is the masterful, correct and magnificent play of volumes brought together in light. Egyptian, Greek and Roman architecture is an architecture of prisms, cubes, and cylinders, of trihedrons and spheres: the Pyramids, the Temple of Luxor, the Parthenon, the Colosseum, Hadrian's Villa. Architectural abstraction has the distinctive and magnificent quality that, while being rooted in brute fact, it spiritualizes it. Brute fact is amenable to ideas only through the order that is projected onto it. Volume and surface are the elements through which architecture manifests itself.

I have seen that this wisdom applies equally well to transforming digital architectures. So, successful managers must apply the wisdom of traditional architectures while transforming digital architectures.

What is the wisdom of traditional architecture that helps modern digital architects? In the glory days of ancient Roman civilization, Marcus Vitruvius Pollio (Vitruvius) wrote about the 'Vitruvius Architecture', penned in his *Ten Books of Architecture*,[11] where he put forth three core tenets of architecture: firmitas, utilitas and venustas. I underline that the three principles regarding digital architectures— architectural vigour (*firmitas*), architectural customer centricity (*utilitas*) and architectural environment synergy (*venustas*)—are crucial for the transformation of digital architecture as well.

Architectural Vigour

Architectural vigour indicates how robust the architecture is. Traditional architecture must withstand many external forces, leading Vitruvius to coin the term firmitas, which represents solidity, robustness, strength or structural integrity and its ability to withstand exposure to the elements. Traditional architectures are exposed to rain, sunshine, wind and other external elements. These could destroy a particular statue, temple or building of any other type you might have built. Similarly, digital architecture should withstand exposure to various adverse events, such as cyberattacks, malware and phishing, among many others. Architectural vigour represents how strong, robust or resistant the digital architecture is to withstand any destruction that may be unleashed by these external forces. You have likely noticed the way the architectural vigour of online websites has been enhanced through multi-factor authentication, which requires you to use multiple methods to verify your identity (OTP via SMS or email, online password, etc.). Further, to enhance the structural stability of systems, organizations have adopted security information and event management systems that help monitor and analyse different threats or security events in real time.

The transformation of architectural vigour has been central to the growth of digital architectures and modern digital architectures are a far cry from the older days of legacy systems (often using COBOL) built over mainframe computers. Often, those systems could not be upgraded easily, were harder to manage and could handle a limited load. Large organizations have faced wide shutdowns due to less robust and standardized legacy systems. In 1994, CISCO had a disaster as its legacy systems environment crashed, forcing the organization to shut down for two days.[12] That was a wake-up call for digital transformation.

In the context of digital architectures, vigour may indicate not just the system's ability to survive an external attack (such as a cyberattack), but also its other technological characteristics, such as robustness, reliability, scalability, flexibility or evolvability. For example, *scalability* is the ability to expand the system's usage by adding resources that may cater to greater computational demands. It may also influence the system's ability to handle a greater workload with minimal ease. The technical characteristics of digital architecture, at large, indicate its vigour and have received heightened attention in the last few years. One example of this is the focus on cloud computing. As a user, you may have seen the technical robustness while using Google Drive, Microsoft's OneDrive or another cloud service. Saving your work files using cloud storage gives you access to a robust and fault-tolerant architecture. So, you can retrieve even the past versions of the files and are assured that you may not lose them—say, due to a disk crash—an event that happened often in the pre-cloud era.

In general, cloud architectures in their various forms—platform as service, software as service, infrastructure as service and so on—are transforming the way organizations create vigorous digital architectures, ensuring high availability, disaster recovery and automated change management, reducing the efforts of organizational technical teams to rapidly detect and mitigate failures (Google, 2022)[13]. Managers must consider various aspects of architectural vigour when thinking about digital architecture. For example, is my web hosting company capable of handling the predicted traffic (and more) while keeping things running smoothly? Similar questions about the technical vigour of digital architectures underline the first class of reasoning any transformation has to consider.

Architectural Customer-Centricity

As organizations embark on the journey of transforming architecture, the concept of customer-centricity becomes increasingly vital. This second aspect is rooted in the Roman architectural philosophy of utilitas. Utility emphasizes that every architecture serves a functional goal. That is, utility ensures that architecture provides some functionalities, say a safe and comfortable place to relax. For example, a house may be intended to provide its inhabitants with a place to get a sound night's sleep. Others, such as commercial malls, serve a utilitarian purpose as well. And this function shapes architectural design. When designing a hospital, for example, one creates a plan that includes emergency exits and speedier routes so that doctors can respond quickly in the event of an emergency. The design of a jail cannot employ a similar concept. There must not be an exit for a simple escape.

The digital architectures follow similar principles. A UPI-based payment system enabling great benefits to the masses through payment transactions across the length and breadth of India connects various stakeholders (say, lenders and loan seekers) across the financial and legal ecosystem of the country. Any transformation must enhance the customer-centricity of the digital architecture. Transformation predominantly revolves around the development of this more customer-centric architecture. For instance, DBS Bank established an analytics competency centre as part of its digital transformation journey.[14] By using analytics, the bank enhanced its customer-centricity to a level unthinkable. For example, voice analytics enabled DBS Bank to identify the reasons for customer complaints and compliments, resulting in a remarkable reduction in complaints and a substantial increase in compliments.

Further, the bank created applications like the Home Connect app, which allowed customers to estimate the cost of a house and monthly EMI instalments. These applications catered to the needs of young and wealthy individuals.

The customer-centricity of digital architecture has evolved over the years. In the 1970s, digital architectures primarily served the purpose of enabling computations for basic decision support. Today, these digital technologies have expanded their functions as they encompass business intelligence and analytics.[15] These may comprise many algorithms, such as classification, correlations, causality, hypothesis testing and similarity matching.[16] These methods and algorithms have the potential to enhance utility, nudging organizations to continuously enhance their digital architectures. Further, architecture transformation may enhance utility for different sets of customers including employees, customers or other stakeholders. For example, the government is building its Digital India architecture for citizens, farmers and so on.

The Digital India programme started an electronic national agricultural market (e-NAM), an online trading platform that provides single-window access to farmers who are interested in selling their produce.[17] e-NAM enables farmers to get better prices as it reduces potential exploitation by powerful middlemen, transportation costs and other such factors. Similarly, the evolving architecture enabling financial transactions, most evidently seen in the growth of UPI payments or digital wallets, enhances customers' abilities to transact or borrow money. As a result, it increases financial inclusion across the citizen population. Similarly, Aadhaar is a public-sector technology that assigns a unique identification number to everyone in India. Aadhaar has simplified identity verification, streamlining various operations across banking and other sectors of the Indian economy. DigiLocker—a Digital India service—allows individuals to securely store and share documents, including driver's licences, mark sheets, passports, Aadhaar cards and so on. These functionalities make it feasible to share these documents electronically at various public service locations, such as airports. Further, start-ups are leveraging these digital architectures—such as Aadhaar and DigiLocker—to provide enhanced utility and rapidly scale their infrastructure. They can leverage underlying technologies to create new applications quickly and respond to evolving customer needs.

Enhancing customer-centricity is a continuous process. So, when undertaking a digital architecture transformation, managers evaluate the customer-centricity hypothesis, assessing how this transformation will enhance utility for their customers—employees, consumers, regulators and so on. That is, a customer-centric digital transformation revolves around understanding and prioritizing the needs, expectations and experiences of various customers. Customers may include consumers, employees, partners and the wider community. To summarize, while making digital architecture customer-centric, one has to think about three related questions:

1. Who are the *customers*?
2. What *state-of-the-art* functionalities are expected by customers or are feasible through modern technologies?
3. What changes in the digital architecture will enhance its customer-centricity?

Architectural Environment Synergy

Often, synergies exist between outside stakeholders and the organization, and digital architectures manifest these synergies. For example, an organization (like Lego) may have a lot of fans who are highly passionate about engaging with the firm and may act as contributors. The concept of venustas, commonly known as 'beauty', serves as the foundation of this synergy. Proposed by Vitruvius, venustas is a concept that relates to how a structure harmonizes aesthetically with its surroundings. Traditional architecture may represent this harmony through captivating architectural and flooring materials or levels of craftsmanship. The architectural theorist Vitruvius identified several characteristics that contribute to such aesthetics, with a particular focus on symmetry and proportion. The aesthetic quality of digital architectures may manifest in the arrangement of different technologies and frameworks or in their synergies with the broader environment. Large feats are achieved and greater performance manifests when the organization harnesses these synergies by transforming its digital architectures.

Open innovations exemplify venustas. Digital architecture transformation may enhance synergies with the environment for broader open innovations. Traditionally, innovation has been closed in nature. Within an organization, all details of innovations are restricted to the people involved with research and development, product development, marketing or other interface departments' staff. These individuals are responsible for conceptualizing and implementing innovative ideas. In the modern age, many organizations are identifying environmental synergies and sourcing ideas from outside. That is, the open innovation approach relies on a lot of external stakeholders, from universities to smaller organizations to collaborative suppliers. The approach represents a paradigm shift in how innovations happen. It emphasizes a greater focus on transparency. Think about how I source the best intelligence, best minds, best effort and best people from outside my organization who might be interested in contributing to the products services, or platforms of an organization. Typically, digital architecture transformation aims to identify and enable these contributions.

A clear example of open innovation is the Android ecosystem. It's an open-source operating system for mobile devices championed by Google, and it was then available to the public under an open source license, which means anybody could put it on their devices. Developers could build their apps and add them to the platform. Due to its open nature, developers were able to easily understand the functioning of the Android platform and build a lot of apps. Beyond Android, non-technology businesses such as P&G and Lego have set up digital architectures for innovation. Lego created the Lego Ideas platform to leverage the power of crowds. Various external contributors created different popular Lego sets, such as the Saturn V rocket, the Ghostbusters and others. Similarly, the P&G Connect+Develop[18] platform allows many external partners to submit their ideas for new products and services. A well-developed digital architecture helps make the connection after analysing the synergies of contributors with internal stakeholders.

The development of open source software (OSS) is an example. The open source movement now boasts over 4,30,000 projects and 3.7 million registered developers.[19] OSS is a special class of software that economizes innovations for organizations. For organizations, open source innovations may lower the costs of digital architecture. The digital architectures of most firms now rely on open source, and some of the software created in open source mode has a key market share in the domain. For example, of all the web servers used for running websites, almost one-third is that of Apache and the other one-third is captured by Nginx—both OSS.[20] However, successfully leveraging the potential of open innovation requires leveraging external contributors' efforts. Open source innovations develop through voluntary contributions, where a founder with a deep interest in developing an idea initiates a project on platforms like SourceForge, leveraging volunteer time and efforts through digital architectures.

To transform and leverage the OSS (by sourcing external contributors), organizations have to think about the basic philosophy of closed versus open. Indeed, it has been a dilemma for many as to why developers contribute voluntarily. Research has examined and identified various internal and external motivations for their contributions. Volunteers contribute for many different motivations, such as learning, sending signals of one's competence, being part of a community and so on. The external participation challenges assumptions about how software may be developed. Eric S. Raymond wrote a famous book, appropriately titled *Cathedral and the Bazaar*.[21] The cathedral indicates a closed source way of developing software and the bazaar is a metaphor that quite appropriately represents the open source paradigm. Should one follow an open source strategy or a closed source strategy for transformation? The answer depends on various aspects related to IP rights, strategy, existing information processors and so on. However, the guiding criteria is just one: does the resulting digital architecture develop high-performance and low-cost capabilities, wherein the costs of IP rights may be included in the overall assessment of the open source approach? Why is open source less costly? Because it relies on creating an aesthetic architecture involving volunteers, the external participants.

Conclusion

Any digital architecture transformation underlines the transformation of vigour, customer-centricity and environmental synergy. Most digital architecture transformations influence all three. Client–server transformation of digital architectures underlines the dynamic. In the late twentieth century, client–server architectures represented a significant transformation of digital architecture that revolutionized organizations. Traditionally, computers and computing devices functioned independently or were interconnected via internal networks. These traditional architectures allowed for printing from multiple computers to a single (shared) printer. However, the advent of client–server architectures brought about a paradigm shift. Via the Internet, organizations could now connect with computers and computing devices located remotely. For example, one of the implications was the separation between client (customer) computers and server (organizational) systems.

These client–server architectures enhanced computing vigour. Client computers required minimal computational power, while the servers possessed substantial computational capabilities. Large technology companies like Facebook, Amazon and Google used servers to execute large-scale computations. Soon, tremendous computational power became readily accessible across the globe via the Internet. Unlike the previous computing era, where a single computer would run software—like Tally for accounting—multiple customer computers could now simultaneously access the software on a server. This model makes it feasible to access news, email, videos and other content over computers or mobile devices. The evolution of client–server architecture commenced in the 1990s and continued to gain greater vigour through technological advancements such as virtualization and containerization. This architectural model has now become indispensable for scalability and security, and the advent of cloud computing marked a further leap in growth.

Beyond enhancing vigour, the client–server architecture enhanced the organization's customer-centricity. Because a large computing and database server could accept and respond to requests from client computers worldwide, the customers had the freedom to connect with the organization any time and from anywhere. That is, an organization's reach among its customers expands exponentially. Customers could access organizations for activities like accessing reviews, reading and commenting on videos, browsing books and other such activities. Further, a centralized server ensures real-time updates on product availability.

Finally, client–server transformation goes beyond vigour and customer-centricity. It underlines the broader relationship between technology and the environment. The development of client–server architectures required the external participation of various technology organizations. These organizations helped design various layers, including presentation layers, database layers and application layers. More complex architectural models are characterized by cloud architectures that represent a new aesthetic dimension. Broadly, the client–server architecture redefined the interconnections between phones, external systems and organizational computing—a synergistic transformation of the relationship between the organization and its internal and external environment.

While transforming digital architectures, managers think about the value of transforming the three aspects: vigour, customer-centricity and environmental synergy. I have outlined a computational view of organization and digital architecture to help assess this value. Notably, each of the aspects may add different organizational values based on how it influences computations. That is, the evolution of the architectural vigour, customer-centricity and environmental synergy transformation requires assessing their computational value.

7

Transforming Work

'It's extraordinary that the world, through the pandemic, was able to work so seamlessly. My dad said something interesting to me—if the pandemic had happened when he was working, it was impossible for him to imagine how it would have played out. In many ways, it shows the potential of what technology has made possible. Having said that, my view is that all this should augment interactions in the physical world and not be a substitute for it. One has seen it in the US, where there has been a clear setback in education and I am sure that's true around the world. Over time, we have to figure out how all of this works to make connections and also what we all experience as humanity and not necessarily be a substitute or be isolating.'

—Sundar Pichai, CEO,
Google and Alphabet Inc.[1]

कर्मण्ये वाधिकारस्ते मा फलेषु कदाचन ।
मा कर्मफलहेतुर्भूर्मा ते सङ्गोऽस्त्वकर्मणि ॥

—Bhagavad Gita

To work is the only right you have but never to the outcomes (fruits) borne through it.
Let not the fruits of your works be your motive, nor let your attachment be to inaction.

The digital transformation of work seems surreal. Consider robotic surgery. The use of robotic arms to conduct surgery is now increasing. One day, when I was teaching digital transformation to medical doctors, they asked me how robotic surgery impacts their professions.[2] Robots performing surgery is a concept whereby the robot gets the inputs sometimes from a doctor sitting afar (and in the future may get these inputs totally through an automated AI system) to operate on a patient. A few days before I taught this class, a friend from New York mentioned a hospital experimenting with robotic surgeries. I realized how the work of physicians or surgeons is being transformed, and this has great potential for saving lives.[3] The lure of robotic surgeries is simple: they may make the doctor, or the hospital, better at treating illnesses by removing geographic constraints. There may still be more work required to make robotic surgeons better than (or as effective as) human surgeons. However, their transformative potential is immense. Due to their wide reach, robotic surgeons will unleash a revolution in the world's physical well-being as more and more people get access to advanced life-saving surgeries for which skilled surgeons are few and there are none nearby.

Throughout the history of medical care, the doctor has needed to be in close physical proximity to the patient in order to diagnose and treat illness. This acts as a geographic barrier that restricts healthcare access for many. Robotic surgeries are not the only way to overcome the geographic barriers. Telemedicine is another healthcare work transformation that has removed the need for proximity, as doctors are increasingly using it to extend their reach. They may diagnose illnesses and prescribe medication to patients via WhatsApp, video conferencing and other related technologies. The case for work transformation was evident in the case of Covid-19. Work-from-home (WFH) is an example of the transformation most organizations engaged in during that time. Imagine the state of humanity if WFH transformation wasn't feasible when the pandemic shook the world for a few years.

Transformation of work is not limited to a crisis (e.g., Covid-19), and it has been a continuous phenomenon throughout the history of human civilization. Often, the advent of new technologies triggers this transformation. Mobility is one example. The work involved in getting from point A to point B has evolved. Until the early twentieth century when automobiles arrived on the scene, people relied on horses for mobility. A horse would be slower and would need to make stops for rest, food and water. It would also be prone to illnesses and injuries.

Besides, managing a horse-based mobility system, such as cleaning up after the horse, was cumbersome. Imagine the magnitude of clean-up required in a city like Delhi (Beijing, New York, Jakarta or Brasília) today if a horse-based wagon system was still the dominant mode of travel. The current petrol- or diesel-based systems of mobility do not face many of these challenges. Travelling on scooters, bikes, cars or buses is usually uninterrupted. However, the transformation continues. An electricity-based mobility system is now being espoused to overcome the limitations of the current system (think pollution).

More recently, the language model ChatGPT has been making waves, with its ability to answer questions or carry out other language-related tasks (formatting, proofreading, software coding, etc.) by using advanced technologies. Specifically, it is powered by AI algorithms on the back end. Such algorithms are able to unravel and enact deep patterns underlying various knowledge works, from software programming to marketing. Today, many people question the rationale or reasons for work. Simply put, the advent of new technologies initiates the phenomenon of work transformation. However, why is it necessary for work to undergo transformation as new technologies emerge? The answer requires us to think about the organization differently.

Organization as a Work System

Previously, I discussed how computations underline an organization. However, an organization is not just an information-processing entity. Instead, it is also a means to structure work activities by tying them to individual routines. The reasoning behind this manifests as work systems. This work system view of the organization helps understand why these routines and activities need to be transformed upon the advent of a new technology. Technology's two inherent relationships with work call for its transformation: complementarities and substitution.

a) Complementarities

The idea of complementarities has its underpinnings in economics research. Formalizing this concept for organizations, two scholars at Stanford University—Paul Milgrom and John Roberts—highlight that complementarities exist between two things when 'doing *more* of one thing *increases* the returns to doing *more* of another'.[4] Realizing greater organizational performance requires work transformation to harness the complementarities between new technologies and elements, such as resources, strategies, structures or managerial endeavours. More importantly, not focusing on these complements leads to losses. For example, consider an electricity-based mobility system. People prefer to use it because it's cleaner.[5] However, the success of electric cars requires a transformation of work, say the transformation of the refuelling (electric charging) process. In turn, this requires reskilling people. It is important to note that complementarities are inherently unseen but real. Many organizations find this out the hard way. General Motors (GM) is an example.

In the 1990s, GM had to transform into lean manufacturing. There was a need to move away from mass production with the advent of computer-based production technologies, such as computer-aided design and computer-aided manufacturing (CAD/CAM) systems. To transform into lean manufacturing, organizations had to realign the activities of human resources, marketing and production strategies.[6] GM failed to do so initially, and as a result, it suffered severe consequences, even after making $80 billion-dollar investments in robotics and other related capital equipment. Complementarities are now gaining prominence due to the advent of AI technologies. To incorporate these technologies, work transformation requires leveraging complementarities.

AI technologies are augmenting humans. However, to tap into their potential, organizations need to transform work. David De Cremer and Garry Kasparov examine this in chess. They underline the role of the best chess combinations, examining how humans augment technologies and vice versa. They emphasize that the best computer program and the best human have rather small effects in comparison to the best process. The human and robot augment each other in the process. De Cremer and Kasparov emphasize the need to work with AI to augment human intelligence.[7] Similarly, a leading technologist, Bill Gates underlines the way to leverage the prowess of digital technologies as he summarizes his assessment of work transformation through ChatGPT:

As computing power gets cheaper, GPT's ability to express ideas will increasingly be like having a white-collar worker available to help you with various tasks. Microsoft describes this as having a co-pilot. Fully incorporated into products like Office, AI will enhance your work—for example by helping with writing emails and managing your inbox . . . Company-wide agents will empower employees in new ways. An agent that understands a particular company will be available for its employees to consult directly and should be part of every meeting so it can answer questions. It can be told to be passive or encouraged to speak up if it has some insight. It will need access to the sales, support, finance, product schedules, and text related to the company. It should read news related to the industry the company is in. I believe that the result will be that employees will become more productive.[8]

To transform work, organizations actively champion complementarities. For example, while implementing voice AI to provide information to their customers, ICICI Lombard complemented the technology implementation with large-scale training of its sales force.[9] At large, when a manager thinks about transforming work, she or he thinks about the complementarities inherent in the various processes, functions, systems or strategies. However, beyond augmenting human beings, many technology-led work transformations automate tasks, often *substituting* humans.

b) Substitution

Work transformation has been about enhancing the effectiveness or efficiency of work systems, and this requires destroying the older ways of working. For example, a reduction in investments in print media may be a fallout of greater use of social media advertising or location-based mobile advertisements. Indeed, many organizations are prioritizing digital advertising and media over traditional media, such as newspapers, magazines or directories (see Table 1).

	Spending (dollars in billions)		Year-to-year per cent change	
	2020	2019	2020 vs 2019	2019 vs 2018
Internet (pure play)	$117.4	$106.4	10.3%	18.8%
TV	57.1	63.8	-10.4	-0.3
Radio	12.0	16.5	-27.2	2.1
Direct mail	11.8	15.7	-24.7	1.2
Magazine	11.3	13.6	-16.7	-4.5
Newspaper	8.7	12.4	-30.1	-11.4
Out-of-home	5.6	7.7	-27.2	8.6
Directories	1.5	2.1	-29.1	-21.7
Cinema	0.1	0.8	-81.6	3.7
Media total	**$225.6**	**$238.9**	**-5.6%**	**6.7%**
Political advertising	13.6	2.1	536.8	-71.3
Total advertising	**$239.2**	**$241.1**	**-0.8%**	**4.2%**

Table 1: Media ad revenue excluding political advertising[10]

Many other forms of substitution are even more stark. Organizations (such as Amazon) are using robots to automate back-end operations. These have led to dark (unlit) warehouses with no humans. Robots working in these warehouses substitute for human workers. This makes the process of stocking and retrieving large pallets more efficient. Similarly, Amazon has pioneered cashierless Go stores, which use AI to scan shopper behaviours and perform automatic checkout without requiring any individual interactions. Not only do these provide convenience to customers, but the data created is a treasure house to understand and improve the retail experience.

These substitutive dynamics are not specific to a particular type of work or workers (say, blue-collar workers). As Bill Gates argues, generative AI influences white-collar workers by creating texts comparable to the ones created by humans, as well as blue-collar workers—through humanoid robots.[11] Indeed, ChatGPT has been the most talked about for white-collar work substitution. Broadly, the large availability of data and analytics is substituting human judgements and decision-making en masse. Data and analytic technologies have been used to propose and test hypotheses at the United Nations and to discern the defensive moves of opponents by the NBA's Houston Rockets.[12] Traditionally, expert humans would be required to make these calls. Even art—such as the creation of opera and music—is not beyond advanced robots.[13] For example, YuMi is a collaborative dual-arm robot. It debuted at the opera, coordinating a team of musicians at the Teatro Verdi in Pisa, Italy. While some substitutions are not urgent, many are motivated by more serious concerns about human suitability for certain tasks.

In other words, many tasks are boring or unsafe for humans to perform, and machines can substitute for human efforts. In daily life, people increasingly prefer an electric toothbrush. A human may find it boring to repeatedly stroke their teeth to clean them. In other contexts, the advent of single-sign-on systems, such as those used by Google and Facebook, allows users to log in to various online websites and services without having to repeat entering their passwords. Similarly, employees log into different organizational work systems (such as ERP systems or portals) using a single sign-on. Such transformation reduces human efforts and makes internal operations efficient. By reducing the repetition of routine activities, organizations unleash employees' creativity and imagination.

Further, new technologies have the potential to substitute for many tasks that may be dangerous. Certain human tasks, such as in mining, scavenging or controlling temperature in steel furnaces, have a strong reason to be automated: they pose safety concerns for employees. For example, the national and international consciousness was under duress for many days, as forty-one miners were stuck for seventeen days in the northern state of Uttarakhand while working on a highway project.[14] Thankfully, they all survived. However, it raises the question: Why should automated machines not substitute risky jobs such as those performed by miners? Finally, substitutions may also be preferred because technologies are less prone to errors. For example, to monitor crops and assess when they need to use fertilizer or pesticide, the use of drones enhances precision, making agriculture more productive. Similarly, in banking and finance, using robotic advisers may enhance the quality of information—about portfolio advice and other financial details—and returns on investments.

Rethinking 'Work'

The presence of complementarities and substitution forms the basis of work transformation. However, managers have to pursue a clear plan. While transforming work, managers must consider whether to substitute or complement existing tasks. Substitution may be possible and not bad at all. However, when strongly manifested and properly leveraged, complementarities facilitate the creation of an advanced work system as well. To choose between the two—complementing or substituting—a manager must think about *what is work*.

There are different ways to think about work, and all of them may be misspecified or short of exact definitions. For example, physicists define work as force times distance.[15] The definition of organization science is more pragmatic. Work may involve a focus on tasks or activities that span individuals. Transforming work entails managing this duality: work is required for an organization, but work is an individual endeavour. That is, while unleashing complementarities or substitutive dynamics, managers must consider the tension between the two: a) enhancing individual prosperity (or limiting any damage to it), and b) creating a more effective work system. The two are resolved by balancing the two aspects of work transformation: transformation of work constitution and work rhythms, respectively.

Work Constitution

Firstly, managers must change the constitution of work in order to transform it. The underlying goal of such a transformation is to create a better work system. Because of complementarities or substitution, the advent of new technology offers the possibility of a better work system. However, organizations carefully transform the work constitution to make the new work system a reality. Tumi, a leading UK retailer that markets high-end luggage and accessories, engaged in one such transformation. Tumi offers customized bags that suit its frequent business travellers' needs (for ease of packing, unpacking and mobility).[16] While the bag quality was good, Tumi wanted to enhance its customer service. To improve its service system, Tumi had to change how it serves customers. It leveraged digital transformation to change the customer service agent's job. They had to become more customer-focused, and the customer could reach them from any of the channels—email, Facebook Messenger or WhatsApp—expecting to continue the conversation where it was left off. Building a human-to-human experience, Tumi transformed how its customer service system worked.[17] The new work system enhanced its likeability among customers.

Work, as a whole, is a carefully constituted entity. A transformation of the work constitution may entail changing the set of activities and tasks or their interdependencies. In 1967, Thompson developed a well-known model classifying different types of tasks. Creating a typology, he outlined three types of tasks: pooled, sequential and reciprocal. In a work system, some tasks may be pooled by nature, with no interdependencies. For example, different call centre agents may attend to calls independently. Tasks that have sequential independence would require the output of one task as the starting input for another. That is, there is a linear relationship in which one task precedes the other. Manufacturing processes, such as an assembly line, typically exhibit this relationship. If you imagine an automobile assembly line, the car flows through the production line as one task is done after another. For example, the doors are first attached to the body of the vehicle, after which the entire car is painted at a subsequent workstation on the assembly line. Finally, more complex reciprocal interdependencies manifest in some other work systems. For example, the design of new products may involve complex interdependencies between marketing, R&D and engineering in determining the functionalities and design of products.

Changing task interdependencies can transform work constitutions. Also, the work constitution may be transformed in other ways, such as by changing task identity, task significance, skill variety required and so on. The purpose of transforming the work constitution continues to be the same—to make a better work system. New technologies offer new possibilities for doing so. Notably, they may call for a reevaluation of the human role. For example, Bill Gates underlines the relationship between AI and humans:

> Although humans are still better than GPT at a lot of things, there are many jobs where these capabilities are not used much. For example, many of the tasks done by a person in sales (digital or phone), service or document handling (like payables, accounting or insurance claim disputes) require decision-making but not the ability to learn continuously. Corporations have training programs for these activities and in most cases, they have a lot of examples of good and bad work. Humans are trained using these data sets, and soon these data sets will also be used to train the AIs that will empower people to do this work more efficiently.[18]

New technologies offer an opportunity for transformation. Especially, modern machine learning and AI-based technologies unravel latent, and surprisingly unique, customer needs. For instance, these technologies could potentially reveal customer buying patterns. Tesco realized the potential of these technologies for creating evolved work for planning and developing offerings. Citing *Wall*

Street Journal reports, Rust and colleagues underline that Tesco used analytics-based technologies (that relied on its Clubcard data) to reveal the need to sell baby wipes and beer together.[19] Since new fathers could not visit bars for their drinks, sales were likely to be higher for both diapers and beer if they were both made available together at local stores. However, leveraging this for superior results required that Tesco locally customize its offerings. Tesco transformed merchandising work so that it could leverage such insights to create superior offerings at the individual level across stores, from hypermarts to neighbourhood shops.

That is, new technologies may not fit with the current work constitution, because technologies differ in terms of their fit with the task. Various theories examine the task–technology fit[20] and related aspects, as captured in the media synchronicity theories[21] and media richness theory.[22] These theories underline how a cutting-edge work system depends on the ability of the manager to reconstitute work in line with the capabilities of the technology. Consider an organization keen to enact dynamic replenishment. Traditionally, replenishment—in organizations such as Walmart, Tesco and Carrefour—had been characterized by fixed schedules, pre-determined routes and historical estimates. Many of these have now moved to dynamic replenishment—a transformation of ways to replenish. Leveraging advanced analytic technologies, organizations harness real-time data from marketing activities, seasonality, weather patterns and other factors. A new constitution of activities helps to dynamically assess demand and schedule replenishments. This also requires changes in restocking and delivery timetables. Further, organizations synchronize coordination across utilization and delivery through advanced technologies for demand forecasting. The new work constitution leads to reduced stock-outs and efficient deliveries.[23] Many organizations have leveraged advanced technologies by carefully transforming the work constitution. For instance, Novartis transformed the work constitution for its sales personnel when it brought iPads to enable the sales team's interactions with physicians. The salespeople may now bring in external experts and influencers to their calls with the physicians. Similarly, Volkswagen USA reconstituted its IT prioritization process to transform which capabilities were developed across the business units.

Work Rhythms

Work transformation entails the changing of organizational work rhythms. The hardest part of transforming work rhythm is recognizing that one exists. Every task has a rhythm to it. And these manifest at the individual level. A rhythm underlines how one feels at work. Within each organization, workers may have different routines based on their specialized jobs, personalities, teams, business conditions and so on. Individuals feel a certain way when they do their job. Organizational work rhythm emerges and may be determined by synchronized coordination across individuals. That is, the organization puts together individuals and their work rhythms are felt by others collectively, creating the organizational rhythm. Think about the organization of traffic on a road. Think about how driving rhythms differ across various towns where you've been. On the road, compare the rhythm of a car versus the rhythm of a two-wheeler. Also, think about the driving rhythm in a big city like New Delhi or Mumbai as opposed to the rhythm in a smaller city or a village. Think about the differences in how drivers accelerate, brake, turn or manoeuvre. I hope this gives you a picture of what rhythm means. Rhythms differ in the way individuals work (say, drive). In organizational settings, for developing software, work rhythms differ across the methodologies being used. Agile methodologies—with scrum meetings, for example— have a very different rhythm from a standard waterfall model for software development. In general business scenarios, quarterly business closings create a work rhythm very different from normal times. The overwhelming push at the end of the quarter makes everyone feel rushed and under pressure.

Transforming work requires changing an individual's job. Various characteristics of a job may change. This changes the rhythm of the organizational work. This is borne out by the Job Characteristic Model (JCM),[24] which has been repeatedly used to demonstrate how job characteristics, such as autonomy or feedback, influence individual feelings of satisfaction.[25] As workers feel differently, work transformations disrupt the rhythms. Managers have to assess and monitor the rhythms carefully, as these influence how individuals feel and how the organization performs. Digital technologies are changing these rhythms, making people feel differently. The use of Google Maps on the road underlines the phenomenon.

Google Maps has transformed the rhythm of drivers and driving. To understand its impacts, imagine yourself driving in a foreign land—a different city, state or country. Without Google Maps, driving rhythm is characterized by uncertainty in navigating, trouble in discerning road signs, finding destinations that one would want to visit and getting there. Imagine the individual feeling if the land has road signs in a language that one does not understand—say, Chinese, Spanish, Hindi or English. The rhythm one feels while making and meeting a lunch or dinner appointment in such a land is marked with anxiety, uncertainty or a lack of comfort. The introduction and use of Google Maps changed the rhythm for the traveller because the mapping app can tell them where traffic jams are and how they can avoid them by taking alternate routes without even asking anyone. This exponentially reduces the uncertainty and related anxiety. Even in one's city, the availability of a mapping app changes the feeling when one knows the exact time they may expect to reach the destination (say, office) for a pre-scheduled meeting. A lot of drivers feeling comfortable on the road change the driving rhythm of the road, in large cities such as Delhi or Mumbai. The drivers are, on average, more relaxed and less stressed.

At large, work transformation makes the work rhythm more soulful. What does the soulfulness of the work rhythm mean? The soulfulness of work rhythms is tacit and less amenable to concrete explanation. To think about soulfulness, think about how you feel when you hear or engage in a symphony or coordinated activity, such as a dance—tango, salsa, bhangra, cha-cha-cha, waltz, samba, flamenco, haka or kabuki. The rhythms on the dance floor for all of these forms differ. However, they are soulful in different manners compared to a novice and untrained person trying to move their hands and legs. In a team, each individual creates (or destroys) the rhythm and is also influenced by the collective rhythm. For instance, when writing a song for a dance, lyricists would experience specific emotions. Singers also feel the rhythm while singing. All of these rhythms are playing for the dancers. Each dancer feels a certain way during the dance sessions and the collective work rhythm emerges, influencing each individual's rhythm. A cycle perpetuates. As different individuals' feelings help synchronize the actions of dancers, so do work rhythms create 'soulful music' in organizations.[26] Any transformation (say, a change of song) changes the soul of the collective rhythm (say, on the dance floor).

Digital transformation, enabling greater use of analytics, is similarly transforming rhythms at work. These transformations have lessened the emphasis on gut feel in board meetings (an unclear computational approach used by managers in the previous era). For example, 1-800-Flowers.com—the world's leading florist and gift company that started with a single shot—relies heavily on data analytics to organize their business. The company's leaders use their intuition to make decisions, combining it with the patterns in the data. The leadership has been quoted as following the maxim: 'In God we trust; all others bring data'.[27] In general, to build advanced rhythms, organizations have to transform how individuals and teams feel when they assimilate more advanced digital technologies.

Specifically, the transformation of work may enhance the soulfulness of its rhythms. Soulfulness may represent how connected the organization is with its customers, employees and other internal and external stakeholders. Broadly, human connections underlie soulfulness. Digital transformation may enhance the soulfulness of work by giving a greater *voice to different participants.* For example, the share of customer's voice within the organization's work may increase. Continental Airlines adopted a data warehousing platform that enhanced its capability to access real-time customer and flight information, helping better understand and meet its passengers' needs and wants.[28] Customers' voices were heard more. In general, a focus on an organization's internal processes (say, serving customers) is invariably transformed, but only when employees feel for customers. For example, through its digital transformation, BloomNet Network enhanced how it internally viewed and related to its customers. The organization developed an analytical ecosystem in the broader floral industry. The company uses the system to transform the work of sending order details to florists and flower growers. The work rhythm becomes more soulful as a human element integrates into the internal processes for exchanging information. Beyond sharing information about the types of flowers ordered, they started sharing customer personas. A customer persona, for example, may be defined as someone who enjoys gifting and uses these gifts to maintain relationships. Such personas offer more meaningful information than mere customer demographics. Sending this information to partners transformed the rhythm of work.[29] The work became more soulful as there was a deeper human connection between customers and florists or flower growers.

Conclusion

Arguably the most important topic in this book, work transformation is challenging human well-being. In fact, work is undergoing such a radical transformation that some are declaring the death of work. This is most clearly apparent in the calls for a universal basic income (UBI). The concept of UBI underlines how advanced technology obliterates the need for human work, suggesting that everyone should receive a monthly fixed income per month without the need to work. The rise of various digital technologies that can carry out human tasks—often in a superior way—supports the argument. Driven by the concern that workers have unfair competition with robots, many countries have experimented with the idea of UBI. Finland conducted one such experiment, initially offering unemployed people a basic assured income of approximately €560 and an assured housing allowance without requiring them to work. This was at a slightly lower level than unemployment benefit, and it led to an increase in employment levels. This is because it boosts individuals' sense of well-being, happiness, cognitive abilities, confidence and trust in themselves.[30] Canada carried out a similar experiment in Ontario.[31] In the US, Andrew Yang, a democratic primary candidate, mooted a similar idea as part of his presidential campaign. Defining it as a 'Freedom Dividend', Andrew promised to offer $1000 a month as a UBI to every American adult above the age of eighteen. [32] He argued that this would help the economy grow faster and increase job creation. While he lost the primary race to Joe Biden (the current US President), the calls for supporting those being influenced are strong.

More importantly, these developments bring to the fore tough questions: How will we think about work? Switzerland politically tested the relevance of work and UBI. Swiss citizens were asked to vote on a proposal to introduce basic income (approximately 2500 Swiss francs) for every adult citizen, regardless of income or employment status. However, people voted against it.[33] The Swiss continued to link work with prosperity, and there are good reasons for people to believe in the relationship. Work is connected to well-being throughout the animal kingdom. Even ants work as they gather food, and some animals hunt to meet their food needs.

The nature of human work is more complex, largely due to the use of advanced technologies. Imagine the food-related supply chains for human beings and compare them with those of animals. Some animals hibernate in winter to balance the body's needs for food with the availability of food. For example, bears work in the summer to store food in their stomachs, large enough to last them through hibernation in the winter. Beavers are known to create dams in the summer where they store food. Digital transformation entails creating complex human work systems. For example, the human food chain is becoming more complex with the advent of blockchains, RosettaNet-type partner integration systems, electronic data interchange (EDI) technologies and many related B2B processes. Understandably, the systems of work are less than perfect, making it imperative for human beings to think about UBI.

However, its use may be desirable in times of emergency—individual or collective. Indeed, both the US and India doled out a large sum of money to sustain the population that could not continue to work during the Covid-19 pandemic. The US sent stimulus cheques multiple times, up to $1200, $600 and $1400 per individual in each wave, for each of the qualifying individuals. India used a different form of stimulus relief. Under the Pradhan Mantri Garib Kalyan Yojana (PM-GKY), the Government of India announced assistance worth Rs 1.7 lakh crore (or $25 billion) for the vulnerable population.[34] The package provided benefits to farmers and rural households. Approximately 70 per cent of these benefits were provided through four major schemes: Pradhan Mantri Ann Vitran Yojana (PM-AVY), Pradhan Mantri Kisan Samman

Nidhi (PM-KISAN), Pradhan Mantri Jan Dhan Yojana (PM-JDY) and Pradhan Mantri Ujjwala Yojana (PM-UY). Indeed, supporting individuals in their time of need is required and may justify breaking the link between work and individual prosperity.

However, there is no need to throw the baby out with the bathwater. Some technology leaders may unintentionally want to do so by proposing a move towards basic income. The transformation of work is not new (as the above example of the human versus animal food chain shows). Throughout history, the conflict has been between the transformation of work to create a more advanced organization and the individual pursuit of prosperity. The tradeoff is resolved when managers think deeply about work transformation. Specifically, they have to think about two broad questions: How will we change the work constitution to incorporate digital technologies? And how will we transform rhythms at work? I have offered some basic principles of work constitution and work rhythm transformation to plan and operationalize such a digital transformation. Thinking clearly about the underlying triggers of the transformations—complementarities and substitution dynamics—will enable successful work transformations.

8

Transforming Governance

'A leader's most important role in any organization is making good judgments—well-informed, wise decisions that produce the desired outcomes. When a leader shows consistently good judgment, little else matters. When he or she shows poor judgment, nothing else matters. Of course, it isn't humanly possible to make the right call every single time. But the most effective leaders make a high percentage of successful judgment calls, at the times when it counts the most.

'Over the course of our lives, each one of us makes thousands of judgment calls. Some are trivial, such as what kind of cereal to buy; some are monumental, such as whom to marry. Our ability to make the right calls has an obvious impact on the quality of our own lives; for leaders, the significance and consequences of judgment calls are magnified exponentially, because they influence the lives and livelihoods of others. In the end, it is a leader's judgment that determines an organization's success or failure.'

—Tichy and Bennis,
Harvard Business Review 2007[1]

Governance transformation is the third element of the DaWoGoMo© model for digital transformation. Governance transformation is an integral part of all digital transformations. Consider the example of Indian software services leader Tata Consultancy Software's (TCS) Secure Borderless Workspaces™ (SBWS™) model. During the Covid-19 pandemic, when many organizations needed to use digital technologies to enable WFH, TCS introduced a transformative operating model framework that enabled employees to work remotely. This framework ensured the continuation of project management practices and systems and enabled the allocation of work, monitoring and reporting, just as seamlessly as in traditional office settings. TCS used the framework to lead a governance transformation across several of its clients. For example, Dutch insurer Achmea sought TCS's assistance in ensuring business continuity while prioritizing employee safety. Working together, Achmea and TCS evaluated different business continuity options and implemented the SBWS™ model for enhancing remote IT equipment's reliability and functionality. Generally, the transformation of governance is the transformation of decision rights. Leading the

DBS Bank digital transformation, CEO Piyush Gupta transformed governance by merging the technology and operations (T&O) divisions. This prioritized decisions by the T&O executives, as they had higher authority in the organizational hierarchy.[2]

Why is it important to transform governance, as carried out by DBS Bank and TCS? To answer, think about decision rights in an organization or at home. Do you allow very young children unconstrained access to social media tools or mobiles? While these tools may help them communicate, parents limit their use. Why? Unlimited use may not be good for children or the family (the organization). That is, parents have the decision right to constrain use. Governance manifests similarly for larger, more complex organizations. In summary, governance transformation requires viewing an organization differently: as a bundle of decision rights.

Organization: A View on What's 'Right'

An organization is a set of decision rights. This is a well-researched and prevalent view in organizational theory. In 1976, M.C. Jenson and W.H. Meckling underlined that an organization may not be treated as a black box and outlined how its governance matters.[3] They combined the logic from the agency perspective—including property rights and finance—to argue how a firm's ownership structure influences its performance. This governance view of an organization outlines the allocation of decision rights as a means to success. Therefore, in addition to changing the work constitution and rhythms, the organization has to reallocate decision rights. While work is innate to humans, so are perceptions about the *'right'* way to work. And, organization is a judgement of the 'right' way to work. That is, it is not ad hoc to conceptualize the organization as a judgement and enforcement of 'what is right'.

Human prosperity stems from our innate ability to judge what is right. Such judgements are hardwired, as they have been crucial for survival and preserving the human physical state. For example, individuals evaluate whether a particular food is right (healthy) to eat or not. Being hungry is judged to be negative and so is eating poisonous food. Further, judgements may differ across individuals. One's understanding of the world shapes these judgements. Therefore, people may have different understandings of the same outcomes. Individuals judge differently as they make sense of information in their world view, often evaluating its emotional effects.[4] Researchers are currently investigating the underlying brain dynamics. Using a tachistoscope, Rolf Gunnar Sandell (1968)[5] conducted a study that exposed participants to two different cigarette brands. These two were associated with varying attributes, such as dullness and stress. The respondents tended to select the brand associated with the attribute that matched their current state. Individuals chose the 'stress' brand when they were experiencing stress. Generally, each individual has different references and judgements. And each one judges the 'right' way to work. In other words, the governance view of the organization underlines that individuals feel differently about the right way to work.

Conflict arises because of the difference in individual judgements (of what is right). At large, it's common for individuals to hold different perspectives and preferences, leading to conflicts at work. Environmental factors and individual tastes, preferences, attitudes and intentions influence choices and lead to conflict. Organizational governance appropriately ascertains *what is right*.[6] So, governance transformation is a way to resolve the conflict arising during the transformation to a digital organization. For instance, when an organization is launching a new product, it faces decisions about whether to pursue a social media strategy, an offline strategy or a combination of both. Individual managers may have a preference for the right combination to choose. A person with authority then has to make the final decision or accept one manager's decision over the other's. Therefore, governance is required to manage conflict. That is, the organization allocates decision-making rights to some people over others throughout the organization.

Allocating decision rights implies that the organization decides whose judgement will prevail. At a practical level, governance may entail determining who decides what and who reports to whom.[7] Organizations do so by determining who has the authority to spend money, to what extent and on what items. To understand this concept better, revisit governance inside a home. In a family setting, various family members each have their own unique roles and responsibilities and a set of rules govern what is deemed right (or wrong). In instances of conflict, someone must decide the right action. Parents often assume this role, particularly when children are very young. Because young children lack the knowledge to judge appropriately, parents' experience and understanding are considered the best judgement of what is 'right'. That is, within a household, the parent's authority resolves conflicts, determining the eventual course of action.

Governance transformation involves the reallocation of decision rights. To transform governance, managers may consider changing the decisions being made, who is responsible for making them or the timing of these decisions. Traditionally, various events call for governance transformation. Some examples include mergers and acquisitions, the launch of a new product line, expansions into a new country, a new strategy and similar others. Over the last few years, digital transformation has emerged as a major event requiring governance transformation. Organizations must transform governance when digitizing. For example, digital transformation may require a new alignment between teams. When DBS Bank transformed digitally, it merged the T&O divisions. This realigned the relationship between technology and business divisions—wealth management, consumer banking and institutional banking.[8] Further, the bank shifted decision rights by elevating T&O roles in reporting structures (reporting directly to the CEO or another manager closer to the CEO). Elevating T&O decision-makers in the organizational hierarchy ensured support for the digitization of processes. However, this digital transformation is often a very thoughtful endeavour. Thoughtful organizations use a well-defined logic for governance transformation.

The Logic for Governance Transformation

The logic for governance transformation underlines the need to assign decision rights to minimize governance costs. The governance cost perspective is linked with transaction cost economics (TCE), which is an academic view of the organization seeing it as a composite of costs associated with various activities. With notable contributions from scholars such as Ronald Coase in the distant past and Nobel laureate Oliver E. Williamson more recently, TCE underlines a central tenet—organizations exist as they offer the most economically efficient way of doing activities. The very act of organizing is predicated on the principle of minimizing transaction costs. In simpler terms, when the associated costs diminish due to the appropriate allocation of decision rights, customers are invariably drawn to the services and products offered by the organization. This phenomenon results in a heightened efficacy of the organization in comparison to its competitors, eventually rendering the latter less effective and, in extreme cases, obsolete. Therefore, managers must think about how to reallocate decision rights to reduce the costs associated with decision-making.

Transforming governance entails a logic to determine how decision rights must be reallocated to reduce costs.

While transforming, managers take into account two types of decision rights: decision control rights (DCRs) and decision management rights (DMRs).[9] DCRs involve determining the decisions to be made and the monitoring of their implementation, such as project selection and progress tracking. On the other hand, DMRs focus on the implementation of decisions, including aspects such as resource allocation and the execution process. Classically, alignment is required between different sets of agents: owners and employees of the organization. Loosely speaking, the former represents the rights of the principals, who determine what needs to be done, while the latter are responsible for determining the 'how' of doing it. The goal of governance transformation is minimizing the costs of determining how to work (work costs) and what work to do (agency costs).

Governance Transformation Reducing Work Costs

Governance aims to reduce costs of work, and this may entail improving work quality. That is, work costs are not short-term monetary costs alone. For example, consider the case of hiring a manager. While a lower-cost manager may seem like a straightforward cost-saving measure in the short term, the cost of potential low-quality work may be high in the long term. Many other types of work costs arise when managers who have decision rights do not make the 'right' decisions. At times, these may lead to even fatal outcomes, as we have seen in the case of some large organizations that have collapsed, such as Satyam Computer Services, which went down due to wrong accounting-related decisions. Finally, work costs may also be high due to missed opportunities. Consider the costs that may arise from the failure to make necessary decisions. Inaction or failure to seize opportunities can be very costly. For instance, decisions against embracing specific technologies or compliance with state policies may have dire consequences for an organization. Similarly, a bank, like the DBS Bank, may suffer if business units do not effectively use digital technologies for decision-making. While transforming governance, organizations strive to minimize the costs related to the conflict at work, by making the 'right' choices.

Reducing work costs involves making the right choices. Hansen characterizes choice as occurring in scenarios involving particular patterns of reactions, including hesitation, exploration of alternatives and uncertainty.[10] While transforming governance, organizations must first consider how digital technologies influence choice-making. Individuals process information to choose between options. Indeed, as March underlines, choosing is a cognitive process that relies on information processing.[11] The empirical support for information processing emerged in studies that relied on pupil dilation and brain-wave measurements.[12] Because it involves evaluating alternative options, choice-making can be engineered. Specifically, technology is becoming increasingly capable of making choices that were previously made by human decision-makers. That is, digital transformation may leverage fast automated decision-making to make the 'right' choice.

The emergence of numerous digital technologies, like the Internet of Things (IoT), illustrates the process of making 'right' decisions. Even for mundane tasks like ordering milk, the right choice may be complex. Ordering, say, one gallon of milk (a common packing unit in many western countries) requires determining the timing of the order. Ordering too soon means one has to deal with the lack of refrigerated space to store it (think inventory carrying costs) and ordering too late will lead to the risk of running out of milk (think stock-out costs). However, it is not easy for humans to continuously monitor the availability of groceries (say, milk) in their refrigerators because of their other engagements. Using IoT, refrigerators can autonomously determine when to order these by self-detecting their levels and previous consumption patterns. Such emerging technologies carry the potential to make the 'right' decisions. Thereby, these reduce the need for human intervention and potential misalignment or conflicts related to grocery orders among family members.

While conflicts in day-to-day chores (like grocery ordering) are relevant, 'right' choices become even more important when decisions carry significant consequences. One such area is surgery, with robots emerging as potential surgeons to challenge the traditional role of human medical practitioners. This raises critical questions about whose decisions should take precedence and this could potentially have life-altering consequences. The ability to make the 'right' choice at the right time during surgeries may be the difference between life and death. Similarly, consider the case of driving. In the US, the biggest cause of death in the age group of five to thirty-four years is motor accidents.[13] Why do these occur? Largely, accidents happen due to errors in human decision-making while driving. Ordinarily, most of us may ignore decisions, thinking them to be trivial. Indeed, many psychologists, in their framing of decision-making theories, underline day-to-day (say, navigation-related) decisions as inconsequential. That is, psychologists may not even consider someone merely navigating around a puddle or walking on a sidewalk as an act of choice, as there is little, if any, conflict in that decision.[14] Nevertheless, these seemingly trivial choices hold paramount importance for designers creating autonomous cars (or using AI-based robots designed for surgery).

Google's Waymo project and certain Indian companies are championing autonomous driving projects, using technologies to make the 'right' choices that save lives.[15] These projects aim to develop vehicles capable of self-driving, eliminating the need for a human driver. Already, autopilot systems have reduced the need for pilot interventions in human aviation. When flying a large plane today, the pilot is not required to continuously control the aircraft's trajectory. Autonomous systems can make the 'right' decisions most of the time (though the pilot still may have the final word). Many modern-day technologies and related decision-making are enabling the discovery of the right decision. In other words, digital transformation is severely influenced by reduced conflict as organizations develop advanced decision-making processes, such as using decision dashboards that facilitate or even automate decision-making. These may quantify and offer objective metrics for decision-making, often removing ambiguity and conflict. Other examples include real-time marketing decisions in social media 'war rooms'. These technologies often enable marketers to take real-time action because they provide objective metrics. Advanced data analytics enables the determination of the 'right' choice in many other domains. Some examples include:

1. Stock trading algorithms rely on predefined criteria and patterns to execute decisions about buying or selling, often within split seconds.
2. For medical diagnosis, applications like IBM Watson analyse medical data, images and patient records to make decisions about illnesses, offering care recommendations that might elude human observers.
3. For credit card shopping, flagging the transactions as being fraudulent (or not) after analysing millions or billions of transactions.
4. Electronic Discovery (eDiscovery) software is used by law firms to decide about the appropriate litigation arguments, by analysing vast amounts of data from previous cases and other relevant sources. Traditionally, this would have required teams of lawyers to painstakingly sift through case files as they evaluated arguments. Using AI, eDiscovery software takes into

account the wording of laws, the preferences of judges and various legal intricacies to come up with an apt and customized argument (or line of reasoning) in a case, specific to the court of a particular judge.

These technologies reduce conflicts and work-related governance costs, as they enable the discovery of close to the 'right' decision. Work costs associated with mistakes or suboptimal decisions are minimized. That is, overall governance costs are lower.

Governance Transformation Reducing Agency Costs

Beyond work costs, governance transformation requires reducing agency costs. *Agency costs* emerge, not due to work but because an individual's interests may not be fully aligned with those of the organization. This misalignment can stem from various factors and the decisions made may not fully align with the organization's objectives. In an ideal world, one might envision zero agency costs. A one-person organization (with a single owner and worker) has perfect alignment. However, reality departs from this ideal scenario. When organizations expand and bring in more individual workers, divergent interests emerge. Agency costs manifest. These costs may manifest as managers offer themselves excessive perks and benefits, which may not align with broader shareholders' interests. It could also mean that workers' decision-making is not transparent. For example, this may lead to misguided investments or mismanagement of funds. At times, workers may seek personal gain over a customer's best interest. For example, a car mechanic may recommend unnecessary services. Agency costs manifest, as these will eventually spoil an organization's reputation by creating bad word-of-mouth (WOM). In real estate transactions, agents might favour sellers over buyers and financial advisers could prioritize investments that yield them higher personal benefits. In the long run, customers lose trust in such organizations. Another aspect of agency costs arises when employees engage in personal activities during work hours, diverting their focus away from organizational tasks. This reduces productivity. During digital transformation, agency costs have far-reaching ramifications. Specifically, they increase governance costs.

The recent use of AI by Hollywood studios exemplifies why. Recently, Hollywood actors and writers went on a strike. Can you guess the reason? Well, among others, the strike was triggered due to differences of opinion about the way to use AI for work in the entertainment industry.[16] Led by the Screen Actors Guild-American Federation of Television and Radio Artists (SAG-AFTRA) and the Writers Guild of America (WGA), the strikers were concerned about the use of ChatGPT and other AI tools in a manner that reduced their incomes and wages. And it was a long strike that crippled entertainment in the US for over four months. The common concern is that ChatGPT may not substitute human workers (also see the previous chapter). This brings conflict, as what is 'right' for writers (workers) may not be good for studio owners. Studios may have benefited more from using ChatGPT. However, in the real world, studio owners are not the only stakeholders. Governments, society, audiences, writers and actors are all crucial stakeholders. Allocating decision rights to manage conflict that may arise if a particular set of choices is not fair or 'right' for some stakeholders is a crucial aspect of governance transformation. The costs of not managing it right are huge (loss of revenue for months for studio owners). That is, despite the tremendous capabilities of technologies to make choices and remove conflict in work, there is a need for governance, as different parties may still conflict because they have different (often competing) interests. The difference in interests leads to agency costs, which are markedly different from work costs.

Trade-offs

The goal of governance transformation is to mitigate the total work costs and agency costs. Traditionally, organizations face a dilemma. Reducing work costs may lead to greater agency costs and vice versa. Let me illustrate with an example. When one hires a competent individual and then monitors them excessively or does not give them the freedom to operate, it may reduce their ability to make sound work judgements. Intelligence and expertise—the very qualities that led to an individual's hiring—cease to be effective. That is, stringent monitoring (controlling employees' work) to mitigate agency costs might increase work costs. Further, as the word spreads, good employees may perceive the organization as not being the best place to work and may not be interested in joining it. The lack of access to a high-quality talent pool may further increase work costs. On the other hand, when one does not specify any guidelines or monitoring, work costs may be reduced. Individuals are creative and empowered in this case. Nevertheless, agency costs may increase as workers may act in self-interest even when it is detrimental to the organization.

However, effective governance transformation reduces this trade-off (see Figure 8). That is, both work and agency costs can be reduced simultaneously. As they enable the determination of the right choices, digital technologies optimize work processes. In other words, algorithms may enhance their ability to work. For example, consider the case of digital transformation for monitoring livestock. These technologies employ sensors and advanced analytics to monitor each animal's vital parameters. As a result, farmers can make real-time decisions to safeguard their livestock's health, reducing work costs associated with delayed responses. Further, consider how the work costs are reduced for a lending organization because of the farmers engaging in this digital transformation. The digital livestock monitoring system may provide objective information for advanced credit approvals. Often, credit approval systems now incorporate numerous parameters, providing more comprehensive insights than traditional lending methods. Now consider the potential of these technologies for reducing agency costs at the lending organization. If employees at the lending organization (say, a bank) act in self-interest and not the interest of the organization, they may deliberately not evaluate a farmer's operations objectively. However, proper governance may reduce information concealment or self-serving actions. For example, digital livestock monitoring may enable the bank's objective assessment of the loan application, furthering the interest of the lending organization. Such a reduction in the tradeoff is commonplace with technologies like blockchain, AI and machine learning. These reveal patterns and trends that human analysis may miss or deliberately ignore, reducing overall governance costs. However, a word of caution is appropriate at this point. A reverse impact of advanced technologies is plausible if governance is not appropriate. That is, if decision rights are not assigned properly, AI may enhance bias in decision-making, as it may further the cognitive limitations in human decision-making.[17]

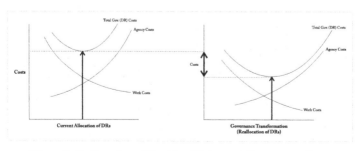

Figure 8: Reallocation of Decision Rights and Governance Costs

Conclusion

In summary, for effective governance transformation, an organization may be viewed as a structure that assigns decision rights and authority to different individuals within the organization. This bundle of decision rights plays a pivotal role in determining which decisions come to fruition and significantly shape the organization's overall performance. Specifically, these influence the overall governance costs, which are a sum of work costs and agency costs. Digital transformation empowers organizations to reduce both work and agency costs simultaneously, aligning them more harmoniously with objectives and the needs of their stakeholders. Several organizations have demonstrated how this can be effectively achieved. For instance, we previously discussed the transformation journey of DBS Bank. DBS Bank's digital transformation revolved around infusing technology into its operations to enhance the customer experience. By merging technology and operations divisions and meticulously considering customer journeys, they optimized work processes, reduced work costs and increased alignment with organizational goals.

The potential to reduce governance costs is increasing as new technologies emerge. For example, technologies such as blockchain, which enables the traceability of goods across various locations, have redefined governance. When Walmart faced a situation involving foodborne illness outbreaks linked to specific suppliers, it turned to blockchain to track its produce accurately. That is, blockchain-based digital transformation enhances transparency by providing a single source of truth. Once recorded, a transaction remains unalterable and visible to all parties. This significantly reduces the need for intermediaries that have traditionally verified the origins of products. Further, smart contracts embedded with decision-making capabilities streamline governance. These contracts can automatically trigger actions and decisions when specific conditions are met, diminishing the costs of managing conflict. However, governance must be decentralized or automated so that decision-making rights to validate transactions are more widely distributed. This leads to flatter organizational hierarchies, reducing the need for numerous intermediaries as data remains current and accurate indefinitely once recorded.

However, new technologies also pose challenges for governance. Consider the scenario when robotic surgeons operate. Who would bear the responsibility for critical decisions that could influence a life? While surgery might be an extreme and complex case, the same dilemma permeates many other technological domains. For example, consider the case of autonomous vehicles. As we move towards self-driving cars, the question of accountability in the event of an accident becomes paramount. Will the blame fall on the software developer, the car owner or other parties involved? Whose decision prevailed? What if an AI-based machine targets civilians for drone attacks? And Bill Gates asks about personal AI agents (reading emails and doing shopping): 'Can an insurance company ask your agent things about you without your permission?'[18] These are complex questions that necessitate consideration of decision rights and the answers may vary depending on legal regulations, societal views and more. Those leading governance transformations need to consider a plethora of possibilities and challenges.

9

Transforming Business Model

'The best way to predict the future is to invent it.'
—Alan Kay,[1] a computer scientist who pioneered
the growth of object-oriented programming and
graphical user interface (GUI) design

The transformation of a business model entails changing how the organization makes money. Numerous technological transformations of business models have now become mainstream. For example, Amazon offers a way for buyers and suppliers to connect through their retail platform; Airbnb enables homeowners to make money from their homes by sharing them for short-term online and Ola (or Uber) enables people to freelance with their cars, offering rides to interested customers through their mobile app. These are innovative digital transformations of business models. So, what is a business model? A business model is a set of interactions and activities that create value. Digital technologies are generating new opportunities to create value. Organizations are transforming their old business models to capture these opportunities. You are likely familiar with some of the new business models through well-known terms: digital platforms, sharing economy models, freemium models and so on.

When you use an app on your phone, you are accessing a digital platform. Android and iOS are well-known digital platforms on your phone. These platforms harness the creativity of millions of developers and the apps created by them are accessible to consumers across the globe. Digital platforms represent the digitization of existing business models. Platforms have been around across the world for a long time, as they bring together buyers and sellers. India's Agricultural Produce Market Committees (APMCs) host *mandis* (markets) across the country. Digitization has led to electronic National Agricultural Markets (eNAM). Broadly, digital platforms represent a more open way of creating and selling products or services. In retail, Amazon and Flipkart have created digital platforms. These enable suppliers to offer products and services. Consumers get an open shopping experience whereby they may access product reviews and compare competing sellers' offerings. The underlying tenet that makes digital platforms so valuable is network effects. That is, once a network of consumers is on board, it becomes very lucrative for potential suppliers to sell on the platform and vice versa—the definition of indirect network effects.[2]

Digital transformation is also revolutionizing sharing models. People have shared for ages. We share public buses, rails, roads and so on. In older days (and perhaps in some places even today), sharing the *tandoor* was common.[3] Typically, the tandoor takes a long time to heat up, and once it is ready to be used, many families may cook on it without any incremental cost to them. This asset (tandoor) utilization creates greater value because of zero marginal costs for an additional user. Similarly, many digital transformations are catalysing the sharing economy to leverage asset utilization. Assets being shared include cars, homes and others. For example, through ride-sharing, drivers offer their cars for rides and potential riders access the data over an Ola (or Uber) app to find them, evaluate them and confirm the ride. Airbnb offers a similar model for sharing one's own house.

New business models are emerging through digital transformation as there is a shift in *what people value*. Also, different stakeholders—customers, employees, partners, regulators, academic and research institutions, investors, community and others—value the set of activities differently. To transform business models, managers carefully analyse digital transformation and its value, considering differences in the assessment of value across *individuals or organizations.*

The Value Logic and Business Model Transformation

Value creation represents a sound logic—one that organizations use to decide how to transform their business model. Pragmatically, value helps the organization act. It forms the guiding criteria, as the organization may be faced with multiple different options for actions.[4] All the other elements of the organization are aligned with the value logic, and decision-makers are given governance authority in a manner that will create more value. For instance, the rising relevance of digital workers has led to the creation of a chief information officer (CIO) (or CDO, CTO or similar position) in many organizations. That is, digital workers have greater decision-making authority in the organization so they may contribute to value creation. Similarly, other elements—say, work systems and digital architectures—are transformed in line with the logic to create more value. Indeed, organizations thrive (and survive) only when they maximize (or avoid a loss of) value. Not surprisingly, this value logic underpins the most widely used theories of organizations, such as the theory of organizational mortality,[5] transaction costs economics (TCE) or the resource-based view (RBV) of the firm. Assessments of value vary across stakeholders.

Further, this assessment is an ongoing process. That is, value is not a static concept. Instead, value transforms dynamically. In the 1980s, the most valuable car for most Indians was the Maruti 800 (or, more formally, the Maruti Suzuki 800). The car had an 800 cc engine. The demand for the car was huge as people transitioned from using two-wheelers—the dominant mode of transportation until then—to four-wheelers. People valued the ability to protect themselves from wind and rain. So, having a roof on top while driving was one of the most valued features. Today, the value that a car offers depends on features such as anti-collision braking systems, lane assist, night vision and so on. Automakers are using the power of semiconductors to offer advanced features to customers, as customers value these features and are willing to pay for them. The business model for automakers has transformed accordingly. Not just automakers, but many large organizations, such as Amazon (for retail) and Uber (for taxis), have championed the most visible and radical business model transformations. These new business models are driven by digital technologies and are assessed as more valuable, especially by customers. For example, the digitization of customer reviews (by online retailers) creates customer value because they get access to feedback that would otherwise be unavailable.

Beyond customer value, digital transformation is also catalysing the transformation of partner-related business models. Research underlines that the digital transformation of partner-related activities creates spillovers, a source of network value. Cheng and Nault unravelled that IT investments upstream (in supplier industries) enhance the downstream industry's performance, underscoring the spillover of value across partners.[6] That is, if the supplier makes greater IT investments, the quality of output and their capability to respond quickly enhance the customers' performance. Broadly speaking, the digital transformation of partner-related value creation is manifesting itself through innovative business models that increase information flows. Vendor-managed inventory (VMI) or 3PL are some innovative business arrangements with partners.[7] Consider Dell. Using digital transformation, Dell worked with FedEx (its 3PL provider) to deliver computers to its customers. Specifically, Dell shared its sales order delivery notes to enable logistics with FedEx. The value of sharing information with a partner is immense.[8] Timely information received from partners increases service levels, reduces demand variability or supply chain costs and enhances inventory performance (say, lead times or inventory levels).[9] At large, the digital transformation of business models creates value by enhancing the quality and nature of connections across partners.

How may managers assess the value that guides business model transformations? They may do so by continuously assessing the value of underlying activities for stakeholders—partners, venture capital funds, banks, regulators and so on. This value may be assessed using different means. However, one simple way is the Net Present Value (NPV) model. This model assesses value over time. For digital transformation, managers may assess NPV as:

NPV = Today's worth of the expected value - Today's worth of expected investments

Doing so, one evaluates value over many periods (say over time t = 0, 1, 2, . . . years) Because value is assessed over time, many Silicon Valley companies have been able to sustain losses over a long period of time, coming up with valuations that may seem enormous and irrational. These organizations sustained losses and kept investing for a greater (positive) value that would be realized only after a long time (say, five or even fifteen years). For example, Uber made its first profits (approximately $400 million) only in Quarter Two of 2023, even when it was founded in 2009. Many such stories are commonplace (e.g., Amazon). However, managers are still faced with a dilemma as they have to look through the hourglass to decide whether the business will be valuable, if not today, in a few years. How do managers make that determination? Netflix's case helps us understand the drivers of value.

Netflix and Twin Criteria for Value Logic

An often talked-about business model transformation is Netflix's rise in the entertainment industry. Specifically, in its initial days, Netflix approached Blockbuster and asked for $50 million to sell itself. The offer was refused and laughed off (as is often quoted).[10] Netflix was in the business of providing online videos and entertainment content (and continues to do so). Blockbuster was the incumbent in that industry. Most people would go to Blockbuster's physical stores to rent DVDs and other entertainment content. Netflix came up with a new digital business model. Early on, Netflix's digital business model entailed bringing the content to customers' homes. At the time, the model was criticized as infeasible and not customer-oriented because Internet speeds were poor. So, even if someone wanted to access an online movie, it would take a very long time to download and watch it. Soon after Netflix launched its model, Internet speeds got better. When the business model became feasible, Netflix was already a significant player. Not only did it become feasible to watch movies online, but consumers also wanted to do so. It was relatively easy for Netflix to become a big player because it was prepared. Many entertainment providers offered content (say, DVDs) through stores. However, they lost their market share very rapidly. Netflix became a huge hit because it had figured out a business model that customers wanted (high valance) and was practically feasible (in the years soon after launch). In general, these are the two criteria that determine the value of a business model: valance and feasibility.

The Valance Hypotheses

A business model is built around the idea of valance. Stakeholders use valance to determine the model's *likeability*, and managers test the hypotheses that a new digital business model will have high valance in the future, if not already. The valance hypotheses of an online entertainment business model, in the case of Netflix, may be assessed by evaluating how much the customers want to view entertainment content online. How much are they willing to pay for it? To what extent will the country's regulators endorse it? And so on. To understand the valance hypotheses, one may reflect on the transformation of retail in India.

Initially, the retail model was centred on haats, which were gatherings of retailers who travelled from afar to serve a collection of villages. People from these villages would walk as well, sometimes kilometres, to access the products they were keen to buy—clothes, grains and so on. Highly liked, haats created positive emotions among consumers. Customers had a positive view of these haats even though they had to walk kilometres, sometimes waiting for days to access the limited options to choose from, for a product, such as clothing for a special occasion like a marriage ceremony. Kirana stores changed the model, and then malls further transformed the retail model. Today, online retailers such as Amazon and Flipkart offer technology apps enabling consumers to shop when they want without stepping outside their homes, and receiving deliveries the next day, within an hour or even minutes (e.g., Blinkit).[11] Why is the online retail business model spreading? Well, because that is what consumers want. In other words, consumers see positive valance in the business model, and hopefully over time, more and more organizations see it as having positive valance. However, managers must assess a second factor, which is the feasibility of a new digital business model.

The Feasibility Hypotheses

The success of digital business models goes beyond positive valance. The digital transformation of business models requires assessing their feasibility. Managers and entrepreneurs question: Is it feasible to enact what is perceived as valuable? Think about the advent of department stores in the West. In the US, the formation of large cities and the rise of railroad networks, which connected these big cities, catalysed the growth of department stores. Subsequently, the introduction of mass-produced automobiles enabled the quick emergence of chain stores such as Walmart and Kmart. With automobiles becoming widespread, it became feasible for people to go to these stores and carry their purchases in their cars over a highway, all on a Sunday. More recently, digital technologies are making it feasible to order items using mobile phones. Players such as Ajio, Flipkart, Amazon and countless others have made it feasible for customers to order online at any point in the day or night and have stuff delivered to their doorstep.

Digital technologies serve as the fundamental basis for many of the current business model transformations, particularly in the sharing economy domain. For example, Airbnb enables people to share their houses. Similarly, Ola and Uber are unleashing value by enabling car sharing. Digital technologies enhance the feasibility of these models; they enable featuring houses to let people assess and choose from available options. In doing so, digital technologies are transforming business models for renting assets. In the domain of digital payments, financial business models are evolving as well. Digital payments through UPI have led to new business models. These manifest in large fintech companies, such as PhonePe, Google Pay and Paytm. There is enormous value in completing instant transactions without having to compute the exact change required. The direct debit from the app creates value as it saves consumers time by not having to travel and withdraw cash from the bank.

Digital transformation is changing the value of various activities, leading to new business models. Think about travel to Mars. Not much is known about customer value at this point. Nevertheless, many visionaries and entrepreneurs are betting on it. At large, the emergence of advanced technologies is enabling many new and valuable functions, features and processes. This is leading to new business models. For example, sensor embedding unlocks value for many products or processes. MIT's stress-monitoring car is an example. It can a) suggest relaxing music and change the voice used in GPS to become smoother; b) vibrate the steering wheel, if the driver seems to be losing attention or consciousness; c) and in extreme cases, researchers at MIT also changed the external colour to send a signal to others about the driver's state.[12] A car with embedded sensors has higher value, as technology makes many functionalities feasible.

Similarly, the digital transformation in insurance is another relevant case. Often, life insurance providers and customers may have different sets of expectations. Japanese insurer Tokio Marine Holdings used digital technologies to offer short-term insurance for specialized purposes such as golf, skiing or even one-day insurance if they borrowed a vehicle from friends or family for a day. This becomes feasible because of digital transformation, which reduces the upfront costs of selling. It enables the company to create offerings on the fly. So, the Japanese insurer uses mobile and location-based technologies to create dynamic offerings, partnering with the telecom partner DoCoMo.[13] In general, the value of business models continues to transform as technologies enhance the valance and feasibility of new features, functions, processes and services.

The Value Divergence Hypotheses

Assessing value is complex because it is not perceived the same by all. When assessing the value of a business model, managers have to think about the perceptual differences across stakeholders. As mentioned before, a *business model is a set of interactions and activities that create value.* Transforming the business model changes interactions to varying extents for different stakeholders. Therefore, their perceptions of valance and feasibility will differ, leading to differences in value assessments across stakeholders. Indeed, a new business model often disrupts value. At any point in time, there is an equilibrium in expectations. Also, the expectations of one stakeholder (say, consumers) are known to other stakeholders (regulators or partners). However, the business model transformation creates a misalignment in understanding, as it takes time to establish one's own expectations and understand those of others. So, managers must don their leaders' hats and think about the multidimensional value impacts of new business models, taking into account different stakeholders. The value is unleashed through digital transformation only if the stakeholders—customer, employee, investor, regulator, community, etc.—are aligned about the feasibility and valance of the new business model.[14]

Conclusion

Business model transformation is an essential driver of most digital transformations. When one is transforming their digital architectures, work or governance, business model transformations offer the guiding light. That is, the transformations are meaningful and valuable only when they are done in line with, and sometimes to enact, a new business model. Transforming the business model makes one focus on value creation through activities and interactions. This may involve thinking about the costs of various transactions. This has deep theoretical foundations.

The transaction cost view of the organization is a cornerstone of management thinking. It helps analyse an organization as the sum of costs of different transactions. Two things determine costs: a) the set of transactions chosen, and b) the cost of the chosen set. To understand, consider the digital transformations of partner-facing operations, such as through blockchain technologies. Network-wide transactions are important value creators in digital environments.[15] The incomplete contract theory explains that inter-firm contracts are rarely complete and there is a cost involved in writing these contracts as completely as possible. Costs involved in specifying contingencies in a contract arise as it is usually infeasible to specify all conditions ex-ante in a contract.[16] Further, the presence of real-time information lowers contracting costs. Therefore, firms may digitally transform by structuring activities using blockchain to reduce the costs of inter-organizational contracting. Blockchain-based interaction may also help specify the contract costs (or remove the need to do so).[17]

In summary, a manager has to think about and evaluate the various business models by assessing the costs of transactions. The most valuable model is the one that minimizes the cost of the set of transactions (or activities) that are most valuable. To assess the value, any manager has to consider two aspects: a) feasibility and b) valance of the business model, and factor in the variations in value across stakeholders.

10

The Extended DaWoGoMo© Model

'By giving [employees] agency and giving them that flexibility and trusting them to do the right thing, I think we will end up again in a net better place where people will be happier and hence more productive.'
—Sundar Pichai, CEO, Google and Alphabet Inc.[1]

How do we implement the DaWoGoMo© model to achieve the desired results? Specifically, we have discussed the DaWoGoMo© model, underlying four different types of transformation: digital architecture, work systems, governance and business model transformations. However, a crucial question is: Are these transformations interlinked and related?

Operationalizing the DaWoGoMo© Model: Additive Synergies

Yes, indeed, they are. Remember, digital transformation is the transformation of the organization. While we have tried to break these into four different types of transformations, a digital transformation entails a holistic understanding and transformation of the organization. Successful managers think about the relationships between digital architecture, work systems, decision rights and business models. Think of the organization as an onion-layer model. At the core are computations enabled by digital architectures, surrounded by different work activities, directed by decision rights and guided by a value logic. The onion-layer model of digital transformation and organization helps explain the context for each of the four transformations.

Each transformation is influenced by one of the other three elements. For example, digital architectures that enable information processing at the core are influenced by how work activities are organized. That is, work activities must be transformed while digital architectures are being transformed. There are alternate ways to transform work. Therefore, conflict arises. To resolve this conflict, there is a layer of governance. The governance layer directs the work transformation through decision rights and authorities. All of these decision rights and authorities are guided by the business model framework at the top. That is, the value framework tells you where the value is, so the governance or decision rights need to be transformed accordingly. At large, the transformation of any element (Da, Wo, Go or Mo) is influenced by the three other elements' transformation. Therefore, when the transformations are executed appropriately at all levels, they unleash synergies. One has to consider the relative effects of transformation at different layers or levels. Think: Which are the levels where you are likely to benefit more? So, how may managers lead a holistic transformation to manage the interdependencies?

Digital leaders may think about an inside-out, an outside-in or a spanning approach. An outside-in approach involves identifying the computations that need to be transformed and then thinking about the transformation of work, governance and the business model. On the other hand, the outside-in approach starts with the business model transformation and then drills down the onion layer, step by step, to the transformation of the digital architecture layer. A spanning approach does not follow a specific direction and may engage in transformations across the board for all four elements. At large, these three approaches may be used iteratively. At different times, an inside-out or outside-in approach may work better, and at other times, the organization may just allow for a spanning approach. A wise approach to drilling down or aggregating at appropriate times will determine success. But, only if you can manage people!

Extended DaWoGoMo© Model: The Cultural Transformation

Is digital transformation just a logical enactment of the DaWoGoMo© model, or do people matter? So far, we have presumed that everyone in the organization wants to transform. However, nothing could be further from this assumption. Resistance follows when you go and enact the transformations of digital architectures, work systems, decision rights or business models. It is unavoidable. Resistance to digital transformation is the norm, largely because people do not just buy into the transformation. In other words, the DaWoGoMo© operates within a wider social context, which may or may not foster a welcoming environment for the transformation.

Much of the organizational research underlines the role of context. In my previous research, I have found environmental dynamism to play a crucial role in a firm's digital strategies. The environment is usually a construct outside the organization. However, the DaWoGoMo© transformation is enacted in a people context, largely within the organization. The concept of organizational culture, climate or atmosphere is more appropriate to think about while implementing the DaWoGoMo© model. I would focus on culture as a characteristic of the broader human context that influences the operational purpose of digital transformation. Culture is sacrosanct to most organizations. Peter Drucker is attributed with having said, 'Culture eats strategy for breakfast.'[2] Indeed, the DaWoGoMo© model may represent an excellent digital transformation strategy. However, unless it takes into account culture, the model only gives limited results. Therefore, I underline that culture influences how well an organization operationalizes its digital transformation. Specifically, I propose an extended DaWoGoMo© model (see Figure 9).

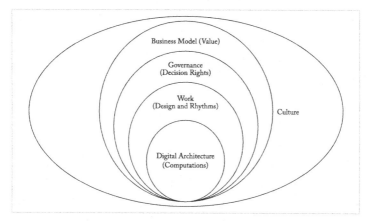

Figure 9: The Extended DaWoGoMo© Model

The extended DaWoGoMo© model emphasizes the role of culture in operationalizing digital transformation. Culture influences resistance or acceptance of the DaWoGoMo© transformation. Any digital transformation depends on how excited different stakeholders within the organization are. It certainly gets into rough places if people are not supportive. The digital transformation at DBS Bank offers an example of the role of culture. While operationalizing its digital transformation, DBS Bank launched the initiative 'Tell Piyush'. Piyush, the CEO, championed this organization-wide initiative whereby people could contribute their suggestions, their concerns or just talk to the CEO. With this initiative, he created an open and participative culture whereby employees could share what was going right, what was not going right or other ideas or suggestions. Over time, the organization created a digital-savvy workforce.[3]

So, what is it that an organization should do to get the culture right? Culture is a very complex construct that has been studied extensively by various experts and researchers. One may work on a sense of enhanced belonging and focus for the employees as part of the digital transformation. Open discussion and training of the DaWoGoMo© model may make the people a part of the organization's operational purpose. Organizations may also think about encouraging creativity, experimentation and learning. A focus on new skill development is paramount. Often, people with older skills may be less valued following a digital transformation, as digital technologies may cannibalize a lot of what they do. Specifically, work substitution (discussed in Chapter 7) may require a focus on new skill development.

Culture may relate to many other aspects. E.H. Schein, who has studied culture, emphasizes that culture manifests at various levels.[4] Based on Schein's arguments, at the surface level, one might see how people are excited or resisting certain transformations, whether they like them or dislike it. They may express this through their refusal to collaborate or their excitement for the changes.

However, managers may be wrong to think about the culture at the surface level. Two layers go below the surface. First, beneath the surface of resistance or excitement, employees conduct an assessment based on their values or beliefs. These values and beliefs help them assess how the transformation influences them. That is, resistance may manifest if employees think that the organizational transformation is really valuable or not. It doesn't add anything to my current work productivity or the way I run the business, or this initiative is not valuable for how we serve customers, and other similar assessments may lead to resistance. Secondly, these assessments typically rely on assumptions. Assumptions could be about how they perceive the managerial team, the organization or digital transformations at large. A positive assumption about the intentions of top managers may lead to excitement about digital transformation. The employees may assume that the managers intend to help them grow or that they are there to exploit them for money. The two assumptions lead to different value assessments and different responses at the surface level to a digital transformation initiative.

In other words, assumptions delve slightly deeper than values (or beliefs) that lie beneath the surface-level reaction to a digital transformation. Further, surface-level behaviours influence values, and this influences assumptions. This implies the existence of a cyclical process. Managers think about breaking the cycle or catalysing it, for cultural transformation. Specifically, many ways may be used to address assumptions or value assessments so that the surface-level responses to digital transformation are positive. These cultural transformations may need to be synchronized with the DaWoGoMo© model transformations to realize the superior effects of digital transformations.

Part 4

The Existential Purpose

11

The Existential Purpose

Life and Digitization

'The human brain has 100 billion neurons, each neuron connected to 10,000 other neurons. Sitting on your shoulders is the most complicated object in the known universe.'

—Michio Kaku, professor of theoretical physics
and co-founder of string field theory (SFT)[1]

Digital technologies are influencing the existence of organizations. Even experienced and well-performing organizations are taken by surprise when they find themselves in a crisis, unsure of their purpose. A fatal outcome often follows. Blockbuster and Toys "R" Us are some leading examples of US companies that have seen the adverse effects of closing down as digital technologies challenged their existence. However, the existential purpose is not just about survival. The existential purpose is a force that may catalyse exponential gains through digital transformation. Indeed, in the last few decades, digital transformations have created phenomenal organizational marvels. In India, the existence of organizations like Paytm, Google Pay or Zerodha have digital technologies at their core. Beyond organizations, even for individuals, identifying existential purpose is an urgent requirement. With the growth of digital technologies, life is changing rapidly, as people's livelihoods, relationships, family, entertainment and minds are transforming radically. What is the existential purpose driving digital transformation?

To answer, one has to think about why we organize and why we are transforming our organizations digitally. Why organize? The answer underlines the existential purpose of organizations and is intertwined with the idea of life. Life is complex and manifests everywhere around us—in animals and humans. It is also evident that animals and humans pursue different existential purposes. Animals are often focused on safety or finding food, especially when they live in a jungle. Humans' existential purpose is more intricate. We purposefully spend our energies and time sending satellites to space (either by building them, funding their creation or celebrating their launch), building and using generative AI applications, making or using autonomous cars, creating and listening to entertaining content, watching or playing sports and so on. In these endeavours, we sometimes make animals our partners. Consider the work done by Maymo, a lemon beagle. The dog's videos are among the most watched, if not the most-watched, on YouTube and these views bring it (or, more accurately, its human managers) the associated returns.[2] Many other animals are getting more attention than they ever got throughout the history of their existence.[3] Arguably, the existence of many of these animals has gained a purpose (at least in some humans' perceptions). The impact of digital transformation on human lives is even more stark and easily relatable. The existential purpose of digital transformation is to help one strategize about and leverage digital technologies more effectively by considering these intricacies.

To unravel the existential purpose, consider this hypothetical dilemma. Imagine you are going through a jungle. Perhaps you are well protected in a heavily armoured vehicle, so no threat may arise to you in any eventuality. And there is no one around. It is a dense jungle and nothing you do can be seen or judged. You have a gun with you, and you notice a lion chasing a deer. The gun offers a perfect shot every time and if you fire a shot, the target will be dead (be very sure, the target will be dead!). The question I ask you is: Will you shoot the lion or not? If you don't shoot, the deer dies, and if you do, the lion dies. Think about the reason you will or will not take the shot. That is, what does an animal's and your own existence mean for you and why? Why must you make a choice? Why do you have the choice when neither the lion nor the deer may be in a position to choose? As I formulate the purpose of existence, think about your answers to these questions.

That is, the answer to these questions helps understand the existential purpose as it outlines the relationship between life, organization and technology (revisit Figure 2 in Chapter 2). In the last few years, it has become clear that any digital transformation requires a deep dive to think: What is life, or more accurately, how do we think about life?

Digital Transformation and Life

Digital technologies have transformed the way we live. My generation grew up without email. It did not exist back then. I remember the first few times I used the Internet to connect virtually to the world. I soon learnt about email (not too long before Sabeer Bhatia made Hotmail a common household name). I continue to use email to date, and I'm sure most, if not all, of you do so too. The advent of email has transformed communication. For modern generations, using digital technologies to communicate their thoughts instantly includes WhatsApp or Instagram. I don't think the modern generation can relate to the world that existed before the Internet. In that world, one had to wait days to communicate or deeply express their thoughts. At family gatherings, people would share pictures of their vacations, graduation ceremonies, birthday parties, newborns in the family and so on. Today, it happens instantly through social media networks. Digital technologies have ushered in the transformation of our lives in a manner that is markedly different from even about fifty years ago. In many ways, digital transformation makes one think about its impacts on *us*. But who is us? More generally, what is life?

Why Should Digital Leaders Think about Life?

Digital technologies are fundamentally transforming how we live, so it is not surprising that digital leaders are deeply thinking about life. The assumptions about life underlie most digital transformations. Even without conscious awareness, a leader maintains a perspective on an individual. She or he may think of the individual as a consumer, citizen, employee, literate (or otherwise), premium buyer and so on. The view of the individual influences how effectively, quickly and responsibly one digitizes. The power of the view of an individual (or life) is enormous. Various theories and models assume a view of the individual. The economic or evolutionary models of life, which form the foundations for various regulatory and managerial choices, define 'who is a wo(man)'. For instance, evolutionary models of life in biology underline that individual behaviours are driven by the gene's characteristics. *The Selfish Gene*,[4] a popular book by Richard Dawkins, underlines that an individual is a means for replication of a gene—a biologist's view of life. Similarly, a popular view of an individual in economics sees him or her as a self-seeking entity. The noted economist Oliver E. Williamson argues opportunism—seeking self-interest with guile—is one of the core characteristics of humans.[5]

Around this view of an individual, a notable debate involving Williamson, S. Ghoshal and P. Moran brought to the fore how what we think about life shapes the actions of executives and others. Ghoshal and Moran challenge Nobel laureate Williamson's view of the individual as an opportunist. *Opportunism* is one of the assumptions about human behaviour that many economists make. Outlined in transaction costs economics, opportunism underlines that every individual is guided by self-interest and may pursue it with guile. Debating with Williamson, Ghoshal and Moran challenge the view, arguing that such a view encourages unethical practices that make the world amoral. They argue that the behavioural assumption is a self-fulfilling prophecy. That is, thinking about other humans as opportunistic leads them—in this case, CEOs, as the theory is often used by these top-level executives for decision-making—to become opportunistic. As an example, they underline that if an individual is targeting to outrun his friend in a jungle when faced with a predator (say, a tiger), we have created a very superficial and amoral work environment. So, what do digital leaders—policymakers, entrepreneurs, C-suite executives, researchers and scientists—think about 'who we are'?

Getting the perspective on life correct is paramount for digital leaders. Specifically, taking an apt view of 'what is life' is crucial to creating a digital transformation that is in line with the broader social ethos and, in turn, shapes this ethos as well. Different, often competing, views of an individual are prevalent. For example, a rational agent model underlines an individual as a perfectly rational being with complete information to make perfect decisions.[6] This model underlies many AI programs. On the other hand, a behavioural economics model argues the opposite. Most popularly seen in the works of Herbert Simon, the model underlines that decision-making is characterized by individual biases and heuristics. So, it is not accurate to presume that an individual is well-organized, has all the information and has the required computational capabilities.[7] Similarly, a game-theoretic model emphasizes that individual decisions are dependent on the decisions of others, as rational agents respond to each other's moves. Society has many different perspectives on life and individuals. Indeed, life is one of the most contemplated topics. Well, it should be. We all have one! So, we develop philosophies to unravel life's purpose.

Most thinking and philosophies outline a *purpose* underlying life. For example, Hindu philosophy may outline life as a cycle of birth, death and rebirth. The pursuit of *dharam, arth, karam, moksh* (freedom from the cycle of life and death) and many other related concepts underline the philosophy. Similarly, Sufi philosophy may underline the purpose of life as a union with the divine through the pursuit of love, devotion and spiritual ecstasy. Buddhist philosophy may see life's purpose as overcoming suffering (to achieve nirvana) and overcoming desires that lead to suffering. Many other philosophies—such as Jain, Christian, Sikh, Ayurvedic and Yogic[8]—define what life is and underline its purpose. Beyond philosophy, scientists also identify life's purpose. For example, as above, evolutionary biologists argue that life's purpose is evolution, and it happens through adaptation and selection.[9] What is life's purpose that guides a leader's approach to digital transformation? The question is tough, as the answer should help identify both life and its purpose *across species* (say, lion, deer and human; revisit the example above). Can we equate all through some conceptualization of *life*?

Neuroscience to the Rescue

As I have found over the years, the modern conceptualization of life—as a neuroscientific entity—is essential to pursuing successful digital transformations. There is an increasing focus on the neuroscientific basis of life, as its links to decision-making are evident across domains such as neuroeconomics.[10] Advanced tools help unravel the neuroscience underlying what humans do, how they do it and why they do it that way.[11] Specifically, advances in techniques such as functional magnetic resonance imaging (fMRI) are helping us understand the neural circuitry that influences behaviours, by analysing activations in the brain. Beyond fMRI, other tools are being used for brain imaging, such as the electroencephalogram (or EEG), positron emission tomography (PET), electrodes that help identify single neuron firings and transcranial magnetic stimulation (TMS) that may influence certain brain areas.[12] Similarly, researchers are also using methods that help assess changes in psychophysical indicators (heart rate, galvanic skin response or GSR, blood pressure or pupil dilation). Like telescopes and microscopes—the tools that helped understand the astronomical space and body—respectively, these tools are unravelling the dynamics within the brain.[13] Can neuroscientific findings help identify what life is?

Reductionism and Reconstructivism

A scientific approach is required to answer the question. Reductionism and reconstructivism are pertinent scientific methods that are renowned in the cognitive neuroscience and computational domain.[14] The methods outline a specific approach that involves breaking the phenomenon into a lower level of analysis (reductionism) and then aggregating it up to develop a holistic view (reconstructivism).[15] For example, in the natural sciences, the reductionist view argues that life manifests in smaller components, such as cells, atoms or biological processes.[16] This approach helped identify molecules as the basis for life. Biologists then unravelled the science behind life based on the basic components: lipids, nucleic acids, carbohydrates and proteins. For example, Francis Crick, the Nobel laureate in medicine, argues, 'So far, everything we have found can be "explained" without effort in terms of the standard bonds of chemistry—the homopolar bond, the van der Waals attraction between non-bonded atoms, the all-important hydrogen bonds and so on.'[17]

Conceptualizing DNA as the basic unit of life is another example of the powerful reductionist approach. However, it is a standard scientific approach to reconstruct after reduction. For example, think about the application of this approach in the natural sciences. While electrons helped represent matter at a reduced level, reconstructivism helped aggregate into atoms revealing the properties of matter in terms of the electron configuration within the atom. In summary, to develop a view of life (or an individual), it helps to first use reductionism and then reconstructivism.

Individual: The Neuroscientific Being

Neuroscience forms the basis for reductionism to conceptualize life. Specifically it underlines life as the state of various neural networks—the network of neurons. The complexity and variations of this state (i.e., life) arise from aspects such as the number of neurons, types of neurons, their connections and so on. Indeed, in the human brain, billions of neurons constitute a neural network. Other species may have hundreds of these. Further, neurons are not all the same. M. Gazzaniga, R. Ivry, R. George and G. Magnum discuss five types of neurons—the motor neuron (spinal cord), pyramidal cell (cortex), association cell (thalamus), Purkinje cell (cerebellum) and somatosensory cell (skin)—in the central and peripheral nervous systems. Connections between the neurons constitute the complex *state* of an individual.[18] A meaningful digital transformation considers the transformation of this state—the transformation of life. However, to do so, an aggregate view of the individual is required at a higher level. To assess the existential purpose of digital transformation, following the standard approach, I aggregate the view of life into brain regions and then into the cognitive schema.

Aggregate View of Life: Brain Regions

In neuroscientific research, *brain region activations* have helped aggregate up, as researchers have used functional magnetic resonance imaging (fMRI) and positron emission tomography (PET) to identify brain activations. Research using these tools has found that individual behaviours are driven by brain region activations, such as regions related to the prefrontal cortex—e.g., dorsolateral prefrontal cortex (DLPFC), orbitofrontal cortex (OFC) or ventromedial prefrontal cortex (VMPFC)—or limbic system (such as the amygdala).[19] While the discovery of new brain regions and their impacts on behaviours is ongoing, a large repository of maps to the brain atlas surfaces and volume has emerged.[20] These tools are catalysing newer identifications that reveal differences in connectivity, function, cortical architecture and the topography of the brain regions. For example, in 2016, Glasser and colleagues used the Human Connectome Project (HCP) for modal magnetic resonance images to identify 180 areas in the brain's cortex, ninety-seven of which were new regions.[21]

It is important to note that brain regions—consciously or unconsciously—run our lives. For example, Gardiner Morse argues, '. . . cortex is an evolutionarily recent invention that plans, deliberates and decides. But not a second goes by that our ancient dog brains aren't conferring with our modern cortexes to influence their choices—for better and for worse—and without us even knowing it.'[22]

These brain regions are known to trigger different behaviours, such as making economic decisions[23] or discerning usefulness and ease of use, and associated with intentions to purchase online.[24] Notably, researchers are identifying various brain regions that engender life processes. For example, professors Angelika Dimoka, Paul A. Pavlou and Fred D. Davis provide a classification of various regions related to the prefrontal cortex—such as the dorsolateral prefrontal cortex (DLPFC), ventromedial prefrontal cortex (VMPFC) or orbitofrontal cortex (OFC)—limbic system (such as the amygdala) and other brain areas. In addition, they link these with life processes—emotional, cognitive, social or decision-making. Specifically, they highlight different constructs (e.g., uncertainty, risk, loss, habits, happiness, fear, disgust, trust or distrust, among others), mapping them to various brain regions.[25] So, how do leaders leverage the neuroscientific foundations to think about the digital transformation of lives? Answer: by aggregating up, to view the individual (or life) as a cognitive schema.

Aggregating Further Up: Cognitive Schema

Students and leaders may reflect on the existential purpose of their digital transformation by assessing its influence on one's cognitive schema and vice versa. Cognitive schema (or cognition) represents a pattern of neural networks (e.g., the state of firings and activations of different brain regions). In other words, cognition represents a higher-level conceptualization of the neural networks. O'Reilly and Munakata underline that human *cognition* manifests as approximately 10 billion biological neurons processing information parallelly. Cognition is a self-governing mechanism. It decides about right and wrong, manages its evolution and even creates digital technologies. Specifically, cognition represents the organization of information in neural networks, and the nature of organization determines information processing for making choices.[26] This view of cognition has evolved over a long time. Barsalou underlined that cognition has been seen as a perception for over 2000 years, and earlier philosophers, including Aristotle and Epicurus in the fourth century and Locke, Hume and Berkeley much later, said cognition is imagistic.[27] Recent advances in programming languages, statistics and logic have transformed this view, and the contemporary view underlines that human information processing involves cognitive schemas that create and store simulations.[28]

At an aggregate level, besides cognition, other similar concepts offer an aggregate view of the neuroscientific being. These include mental models, frames and knowledge structures.[29] Cognitive framework is a broad term often used for these aggregate views of the individual. Also, the similarity between these related concepts has been outlined:[30] '. . . cognitive frameworks, also referred to as mental models, knowledge structures, scripts, and frames are closely tied to how individuals make sense of and act within their environment.'[31]

It is imperative to note that cognition is not a theoretical concept. It is very accurately measured as a knowledge structure. In other words, cognition represents knowledge. Rumelhart argues that cognitive schemas represent knowledge spanning ideologies, cultural truths, the meaning of words and so on. In general, schemata of cognition represent knowledge at different levels of experience and abstraction.[32] For example, it may represent relationships between concepts. Also, it is often complicated and entangled. Further, a cognitive schema may also represent ways to manipulate these representations. Cognition can discern what to look for around it, when to look for it, how to add new information to itself and how to learn new concepts. Using this conceptualization, cognition as a knowledge structure is objectively measurable. Various methods used for measuring knowledge structures include multidimensional scaling and cognitive mapping.[33] Different methods, such as the repertory grid and free elicitation, are used to measure various aspects of cognitive knowledge structures, such as dimensionality, abstraction and articulation.[34]

Conclusion: Digital Transformation and Cognition

Digital transformation's existential purpose is to transform our cognitive schema (cognition). As individuals, we are strongly driven by a desire to create and harness our cognition. To do so, we classify objects, events and actions to discriminate between them. Schemas help us organize the world in our minds. These are such powerful forms of life that we impose schemas on random entities. For example, people are known to see faces in random images—a phenomenon called *pareidolia*. As it may be, it is known that a face is often perceived in the photographs of Cydonia Mensae (on planet Mars) taken by the Viking Orbiter in 1976. Even when the Mars Global Surveyor in 1998 took close-up pictures showing dunes and hills, people liked to perceive that they saw faces.[35] To unravel the existential purpose, it becomes imperative to unravel: What is the impact of digital technologies on cognitive schema?[36]

However, it may not be accurate to think that the effect is only one-way. The cognitive schema may also transform organizations and digital technologies. That is, in a cyclical process, cognition transforms organizations and technologies, which in turn transform cognition.[37] Existential purpose underlines the cycle. So, we think about the digital transformation's existential purpose by thinking about its impacts on the organization and cognitive schema and vice versa. Next, I unravel the existential purpose underlying *what* the organization digitally transforms. In other words, what is the existential purpose for all living beings that makes creating digital capabilities (the instrumental purpose of digital transformation) a strong force for success?

12

Revisiting the Instrumental Purpose

Digital Transformation and Individuals

'There's a conflict between what we want to do and what we're actually capable of doing.'
—Dr Adam Gazzaley, David Dolby Distinguished Professor in neurology, physiology and psychiatry, University of California, San Francisco [1]

'I see AI and machine learning as augmenting human cognition a la Douglas Engelbart. There will be abuses and bugs, some harmful, so we need to be thoughtful about how these technologies are implemented and used, but, on the whole, I see these as constructive.'
—Vinton Gray Cerf, widely regarded as the father of the Internet[2]

What is the reason building digital capabilities unleashes a force that propels the organization to higher performance? The instrumental purpose to build digital capabilities is effective when it catalyses the existential purpose of life (evolution of cognitive schema). The cognitive schema (life) seeks digital capabilities to pursue its emotional goals. This is because, instrumentally, we exist to realize positive emotions.

Emotions and Existence

The pursuit of emotional goals is the force that drives human beings (and other species) to act. This force is rooted in the cognitive schema. A cognitive schema represents life and has its own *will*. It may aspire to create products, run organizations, write manuscripts, conduct experiments and take care of oneself, family, community and society. Sometimes, it may do the opposite—tear up manuscripts, not take care of oneself, be destructive and so on. Moreover, cognitive schemas are not unique to humans alone. Animals have cognition as well, albeit not as advanced as that of humans. Therefore, we can understand the existential purpose driving life by examining the existential pursuits of animals. What existential purpose do animals pursue?

Previously, complex cognitive mapping and analysis among animals revealed that the existential purpose is *the pursuit of emotional goals*. For example, studying Clark's nutcrackers (*Nucifraga columbiana*), Kamil and Cheng found that they search very precisely for food as they identify and link various landmarks.[3] Is there a hidden, latent force that guides this search? Alternatively, what drives animals to act? Research provides evidence of the latent force, revealing that emotions wire and drive pigeons. Specifically, changes in respiration patterns are observed as pigeons experience emotions triggered by threats or food availability.[4] At large, emotions drive various pursuits, such as selection, planning, action and feelings. Existentially, the pursuit of emotional goals underlines most endeavours, even in humans. While we act to preserve or create emotions, the human existential purpose goes beyond the food search.[5]

To realize positive emotions, we aspire to have a feeling of health through a disease-free body; feel beauty by listening to a favourite song or watching a dancer move across the hall; feel the rhythm when our star player makes an iconic throw, plays a cover drive, hits the beautiful backhand or makes an awesome flip kick; feel aesthetic by viewing a painter's masterpiece or get excited while watching a match ending in a tense game. Going beyond survival is not restricted to humans. Even animals are known to be driven by an existential purpose that goes beyond the self. For example, capuchin monkeys (*Cebus apella*) have been found to demonstrate markedly significant prosocial behaviours.[6] Experiments using a limited-form dictator game show that these monkeys demonstrate a penchant for giving food to other monkeys. Monkeys shared food even when there was a zero-sum game, i.e., when they ended up gaining less for themselves because of giving. In summary, the existential purpose of individuals is driven by emotions and these often go beyond the self.

Much of an individual's existential purpose is the pursuit of complex emotions, often related to phenomena outside of us (just as was the case with capuchin monkeys). That is, throughout one's existence, one's purpose involves thinking about (one's own and others') emotions—joy, comfort, pain, anxiety, empathy, love, self-awareness, pride, arrogance and many others. Emotions may embody the essence of a human being. Underscoring the role of emotions, Joseph LeDoux argues, 'Emotions, after all, are the threads that hold mental life together. They define who we are in our own mind's eye as well as in the eyes of others. What could be more important to understand about the brain than the way it makes us happy, sad, afraid, disgusted, or delighted?'[7]

Even a less-than-year-old child understands discomfort when hungry, even when they do not know anything about the world—do not recognize vehicles (buses or trucks) or relationships (who are mom and dad?). Various emotions (joy, fear, etc.) are known to the child as they cry or laugh, even when they may not know much about the factors causing them. As they grow, the child forms a deep relationship with the primary caregivers. Over time, as they interact more with society (parents and families, to begin with), the child develops a deeper understanding of life, understanding empathy and other emotions.[8] Over time, the child acquires self-awareness. Not having much knowledge of the world, the child (and most other individuals) identifies life's existential purpose through the prism of emotions, pursuing activities to achieve desired emotional states. The brain's evolution has a deep message for digital leaders. Life's existential purpose is to experience (or avoid experiencing) certain emotions—to feel a certain way.[9] Throughout one's existence, life evolves as one develops a more nuanced understanding of the emotions of oneself and others. Therefore, to understand the rationale behind building digital capabilities and unleashing the power of instrumental purpose, leaders must grasp the significance of emotions in our lives.

Defining Emotions

Understanding emotions has been a complex endeavour. Various researchers have mapped emotions. Through an Emotion Profile Index, Plutchik mapped the emotions underlying the evolutionary process. The index consists of sixty-two forced-choice emotion descriptor pairs that measure eight basic emotions. Basic emotions are the ones that are core to human beings. Other emotional typologies conceptualize emotions differently.[10] For example, some identify three aspects of underlying emotions: pleasure, arousal and dominance.[11] Despite these typologies, it would be wrong to think emotions are well understood or understood at all. Defining emotions has been a problem for centuries.[12] Klaus R. Scherer[13] cogently summarizes the concern: 'Even though the term is used very frequently, to the point of being extremely fashionable these days, the question "What is an emotion?" rarely generates the same answer from different individuals, scientists or laymen alike.'

So, how do we identify an emotion? Bechara Antoine and Damasio Antonio argue that emotions represent changes in the body's (somatic) state in response to certain stimuli, and such changes may be observable to outsiders (e.g., heart rate, endocrine release, facial expressions, changes in posture or behaviours, such as freezing, among others).[14] Thus, they define emotions as feelings that may involve the central nervous system (CNS). The CNS is responsible for releasing various neurotransmitters (such as dopamine, serotonin and noradrenaline), changing the state of various regions of somata or changing the transmission of signals from the body to the somatory parts. Scientists have documented the link between emotions and bodily changes. Changes in the somatic state (the state of internal systems such as respiratory, musculoskeletal, endocrine, digestive or circulatory systems). This may manifest as hormone secretion or a change in heart rate, say, when one comes face-to-face with a threatening animal (stimulus).[15] Similarly, a person feeling embarrassed may blush, and another feeling afraid may experience a racing heart (much like respiratory patterns in pigeons, discussed above). In summary, when an individual perceives a stimulus (say, a threatening animal), the feeling of the body changes (say, an increase in heart rate) due to the emotion engendered (e.g., fear). And individuals prefer to experience certain emotions (body states).

Digital technologies are influencing emotions, and this is silently transforming our evolution. Gary Small and Gigi Vorgan argue:[16]

> The current explosion of digital technology not only is changing the way we live and communicate but is rapidly and profoundly altering our brains. Daily exposure to high technology—computers, smart phones, video games, search engines . . . stimulates brain cell alteration and neurotransmitter release, gradually strengthening new neural pathways in our brains while weakening old ones. Because of the current technological revolution, our brains are evolving right now—at a speed like never before.

To examine the role of digital transformation in human emotional evolution, a look at the cognitive schema's neuroscientific dynamics is apt.

Emotions and Cognitive Schema: The Neuroscientific Connection

Indeed, emotions have a strong neuroscientific basis. People's emotions—hate, joy, love, anger, disgust, satisfaction and others—originate in areas of the brain that are distinct and different. A very large body of research underlines the neural circuitry that engenders emotions.[17] The brain-split experiments demonstrate that a limbic system is linked to emotion creation.[18] The limbic system is proposed to be composed of various parts, including the hippocampus and dorsolateral prefrontal cortex (DLPFC), amygdala, medial frontal gyrus, anterior cingulate cortex, posterior cingulate cortex, orbitofrontal cortex and ventral striatum.[19] Broadly speaking, the limbic system comprises a set of brain structures that help regulate emotions.[20]

With a neuroscientific basis, emotions are widely mapped to activations in different brain regions.[21] For example, Joshua D. Greene and colleagues compared activations across brain areas for tasks associated with different types of morality[22] and found greater activity for the moral personal task (than the moral-impersonal and the non-moral tasks). Further, this activity was found in certain parts of the brain responsible for generating emotions, such as the medial portions of Brodmann's Areas, including BA 39 (angular gyrus, bilateral), BA 31 (posterior cingulate gyrus) and BA 9 and BA 10 (medial frontal gyrus). Neural correlates have been identified for even the most subjective emotions, such as those associated with the sense of aesthetics. For example, one may experience aesthetic wellness due to an experience of visual, musical, mathematical or moral beauty. A common emotional neurocircuit—field A1 of the medial orbitofrontal cortex (mOFC)—underlines the experience. Similarly, neuroscientific activations underline the evaluations of a piece of cake and a piece of music.[23] A positron emission tomography (PET)-based analysis reveals that pleasant (or unpleasant) feelings arise through musical dissonance, associated with activation in paralimbic brain regions. Further, a pleasant music experience, often defined as 'shivers-down-the-spine' or 'chills', enhances cerebral blood flow in the ventral striatum, orbitofrontal cortex, midbrain, amygdala and ventral medial prefrontal cortex—circuits that are associated with rewards.[24]

In summary, different brain regions help evaluate stimuli to create emotions. Richard J. Davidson showed that different brain areas perform different functions and there are individual differences in realizing emotions due to this physiological asymmetry.[25] This implies that not all of us can experience similar emotions, i.e., it may not be physiologically plausible for everyone to experience the same emotions. Nonetheless, we are hardwired to act *emotionally*.

Emotions, Evolution and Digital Transformation

Realizing emotions unleashes the existential force for digital transformation because that is how life evolves. Emotions are linked with evolution. Specifically, emotions are a product of evolutionary wisdom that has accumulated a lot of intelligence for the growth and well-being of individuals.[26] For example, emotions provide us with a signal to act. In the natural world, survival requires physical fitness, and evolutionary biologists underline that humans have to act to survive and thrive.[27] Even when emotions may have negative consequences, they are often a correct evolutionary response. For example, Aristotle argued that anger is a perfectly rational response to an insult as it helps identify unpleasantness and pain, and others support this view.[28]

Such is the force of emotions that individuals even *reappraise* them. That is, individuals tend to emote in a manner that is not based on real events but on what we feel is the *right* emotion to experience. We are so wired to feel a certain way that we anchor our lives to expected emotional states. Indeed, research has shown that individuals cognitively tend to transform their emotional experience, reappraising adverse events to reduce negative affect (though this may have physiological, cognitive or social costs).[29] For example, we don't want to feel sad (in order to avoid crying) when we are with friends. Similarly, we may not show we are afraid (even when we are dead scared) and so on. Neuroscientific research has unravelled such emotional reappraisal, outlining anchoring. Which means our brain's neuroscientific apparatus is hardwired to help us evolutionarily seek anchored desired states, even if it means doing so through reappraisals. Using an fMRI study, Kevin N. Ochsner, Silvia Bunge, James J. Gross and John D.E. Gabrieli find that the prefrontal cortex plays an important role in helping individuals reappraise emotions.[30] Neuroscientifically, it may not be very far-fetched to conclude that we are wired to act in a specific manner because of emotions.

The reason emotions guide us to digitally transform in a certain manner is because of evolutionary dynamics. Certain emotional states aid in evolution. This is in contrast to natural evolution that happens over generations. Natural (physical) evolution is not the only, or preferred, approach in the digital age, where there is little time for genetic mutations (which happen over centuries).[31] Our external circumstances do not force us to choose. Instead, one uses emotions to make choices that lead to a better lifestyle.[32] Realizing positive emotions requires computation, as does the natural evolution process. Evolutionary biologists Stuart A. Kauffman and Edward D. Weinberger (1989) highlight the relationship: 'Adaptive evolution is, to a large extent, a complex combinatorial optimization process. Such processes can be characterized as "uphill walks on rugged fitness landscapes".'[33]

Therefore, a force is unleashed if we pursue an instrumental purpose to leverage digital technologies to realize preferred emotions. Such a force offers an existential advantage. Therefore, to succeed in their endeavours, organizations pursue a clear instrumental purpose for digital transformation.

Revisiting the Instrumental Purpose: Why We Do 'What We Do'

Part 1 of the book underlines that the instrumental purpose of digital transformation is the creation of digital capabilities. Because life's existential purpose is emotional in nature, digital capabilities that help realize individual emotional goals are crucial for success. They unleash an existential force that drives success. Life's (cognitive schema's) emotional goals, across species, may be met through these digital capabilities. Think about how emotions are created—by appraising stimuli (see above). Digital capabilities act as stimuli (or counter-stimuli), catalysing and speeding up the realization of emotions. Emotions are a response to stimuli. Revisit the story in Chapter 11 about the chase in the jungle. The deer may be scared to see the lion (the stimulus). Having a protective vehicle around oneself makes the human feel safe, as the stimuli are now the vehicle and not just the lion (think about the human without the vehicle or the gun). The stimulus-response process through which emotions are engendered forms the foundation of life's evolution. Digital technologies—hardware and software—enable the capabilities required to create stimuli (the car, the gun and so on) rendering emotions and transforming our lives.

That is, one usually has a reference point that is used to compare one's position and assess the emotional response.[34] Emotions are also engendered through the appraisal of progress towards the goal. Charles Carver S. and Michael F. Scheier[35] present a model based on a feedback loop that explains the engendered emotion, or the affect. They outline a secondary system that monitors the progress of an action-based system working to achieve the desired goals. When the progress towards meeting the goal is faster than expected, individuals perceive a positive affect and a negative affect arises when the progress is slower.[36] Similarly, satisfaction was found to be even more influenced by the velocity with which the goals are being pursued instead of the position of goal fulfilment.[37] Therefore, organizations' instrumental purpose—to develop advanced digital capabilities—shifts the speed for creating positive emotions. For example, ex-IBM CEO Gini Rometty underlines that data sharing through social media will lead to a greater focus on oneself. And organizations will be able to develop capabilities to better relate to the customers.

'If you have a call centre, it's no longer about a script', she emphasizes, 'it's about a dialogue.'[38]

Similarly, Sherry Turkle, the Abby Rockefeller Mauzé Professor in science, technology and society at MIT, argues that digital transformation may even transform how humans emotionally act against self. A digital machine may measure the emotional state of a corporate vice president, say, by assessing her or his galvanic skin response or pupil dilation very precisely and non-invasively, even when the person may try and hide their emotions.

The machine might say, 'Mary, you are very tense this morning. It is not good for the organization for you to be doing X right now. Why don't you try Y?'[39]

Digital transformation's instrumental purpose becomes crucial as the evolution of life (cognitive schema) requires capabilities. However, humans have access to a very limited set of *capabilities*. We are not the fastest runners or swimmers, and we can't fly. In the animal kingdom, birds can fly. We felt a deep desire to acquire these capabilities, as they enhanced our emotional experience as a species. Therefore, digital capabilities link with human emotions. This is evident in the soaring valuations of iconic digital organizations. Facebook, a social networking platform valued at close to a trillion dollars,[40] enables social interactions and meets the emotional goals of over 3 billion active monthly users.[41] Individuals use the digital platform to realize various emotions such as sadness, anger, love, frustration and connections, among others. Facebook is not the only digital company to catalyse individual emotional goals. The digital phenomenon extends beyond the use of free social media apps and websites.

The realization of emotional goals underlies a near-global acceptance of the Internet, mobile and other related technologies. Approximately 64.6 per cent of the global population (5.18 billion people) uses the Internet and approximately 4.6 billion use social media.[42] People just don't visit social media; they spend a large amount of time on these websites and apps—an average of 151 minutes a day, as of 2022.[43] Many other digital technologies, such as those created by Google's YouTube, Twitter (X) and WordPress, enable individuals to realize their emotional goals by enhancing their abilities to create or watch videos, send and receive messages (140 characters or less) and write or read blogs. In summary, the widespread acceptance and use (leading to enormous valuations) of these digital organizations underline the value individuals place on *digital capabilities*. Each user's cognitive schema contributes to the valuation as it assesses the effects of digital capabilities on (positive) emotions.

Therefore, a more vigorous force is unleashed when the instrumental purpose of digital transformation (digital capabilities) creates the most valuable emotions. An example of this is the digital transformation that occurs during times of existential crisis. Consider the digital transformation by Médecins Sans Frontières (Doctors Without Borders). Médecins Sans Frontières is a humanitarian organization that provides medical assistance in crisis-stricken regions globally. They harness digital technologies for providing a vital tool, offering telemedicine and remote healthcare delivery, thereby enhancing emergency medical care in areas affected by conflict, epidemics and natural disasters. Médecins Sans Frontières brings hope and relief to those facing life-threatening situations. Similarly, the exponential acceptance of digital transformation during Covid-19 underpins the fact that digital capabilities unleash the existential force for life's transformation. As argued by Sundar Pichai, the current CEO of Google:

> The global pandemic has supercharged the adoption of digital tools. Digital payments, for example, have enabled families across India to access goods and services during lockdowns. For them, grocery delivery services have been invaluable—though I'm sure my grandmother misses haggling over the price of her vegetables in person.[44]

Indeed, an organization is meant to enhance our emotional experience. For example, if we compare driving a luxury car (say Mercedes) with driving a Maruti Suzuki (in India), the technical sophistication, engine power, suspension, etc., differ, leading to a different quality of driving experience.

The digital transformation of the car (say, to include features such as lane assist) is a pursuit to enhance our emotional experience. Perhaps nothing else underlines the power of the instrumental purpose more than the efforts to offer various citizen services digitally. Individuals want to feel sure, certain and hopeful for the future. And the digital transformation of government services offers capabilities to fulfil these wants.

Digital capabilities for accessing essential government services enhance citizens' convenience. Anxiety, angst and discomfort are reduced as citizens easily access crucial services online. For example, in India, digital payment systems built using UPI and Aadhaar have provided capabilities for financial inclusion and cashless transactions. Similarly, mobile-based banking unleashed economic empowerment for the underprivileged. Many people excluded from the financial system wished for dignity while getting loans, a feeling of accomplishment and pride in transacting with ease, and satisfaction in receiving fair credit terms. Online banking helps many realize positive emotions associated with these goals. Not surprisingly, across the world, there is an increase in digital payment systems for making payments. For utilities (water, gas, electricity, etc.) Rs 145 million was paid digitally in 2020, up from Rs 41 million in 2014; those submitting taxes online went up to 143 million in 2020 from 73 million in 2014.

India's Digital India campaign has created many capabilities through initiatives such as common service centres (CSCs), digital payment systems, digital literacy programmes, e-governance services and Matr Shakti. Scores of underprivileged citizens could never realize the positive emotions that are now enabled by these capabilities. For example, consider the case of common service centres (CSCs). For the longest time, many citizens have been deprived of access to digital technologies, and the digital divide prevents access to crucial services. Physical centres that serve as digital hubs and CSCs provide citizens with access to government services, especially for those in rural and remote areas. These services may include enrolling in Aadhaar, digital payments and information on government schemes. Easy access to these makes people feel included. Similarly, during travel, a Digiyatra app eases the security checks. Not surprisingly, there is an exponential increase in the government's digital transformation. Across the world, the number of applications online has surged: for birth certificates, from 44 million in 2014 to a staggering 156 million in 2022; for driver's licences, from 29 million in 2014 to 146 million in 2022; for identifying cars, from 27 million in 2014 to 150 million in 2022.[45] In summary, developing digital capabilities has made many realize positive emotions.

That is, the instrumental purpose of organization and the existential purpose of life are interlinked. An organization that creates digital capabilities enables the cognitive schema to evolve by realizing the desired emotional state, the existential pursuit across species.

Conclusion

This chapter underlines the role of emotional goals in driving digital transformation. An organization's digital transformation is driven by individual emotional goals, which lead to the manifestation of its instrumental purpose. Much of digital transformation's existential purpose is to evoke positive emotions in our day-to-day lives. The organizational instrumental purpose of digital transformation—its digital capabilities—is guided by this existential drive. Known for its innovation and commitment to making information universally accessible, Google offers digital capabilities for many positive emotions. For instance, the use of Google Maps has reduced people's stress and confusion during travel. Similarly, searching for many products and services is now instant with Google Search, reducing anxiety for many. Aside from that, emotional evolution has just begun. Endless possibilities to develop advanced digital capabilities lie ahead, but only if we can harness them to catalyse human evolution. Sundar Pichai, CEO of Google and Alphabet Inc., may help you reflect as he thinks about the potential of Google's Calendar technology: 'I always think about why doesn't Google Calendar tell me, "Hey, it's a nice day outside. You haven't walked . . ."'[46]

Furthermore, Pichai wonders if it would be beneficial to have calendars suggest dinner with family or help someone make the time to learn Spanish. Pichai is not alone in suggesting linkages between digital capabilities and emotions. Bill Gates underlines a similar connection when he asked an AI model to write a response to a father with a sick child: 'It wrote a thoughtful answer that was probably better than most of us in the room would have given.' [47]

Emotions are universal across species. Emotions are widely prevalent across animals as well. However, humans are distinct. Beyond the daily chores of life, the human cognitive schema has the power to judge the complex, the philosophical and the unknown. Not surprisingly, the evolution of human emotions involves higher-order emotions, which are beyond one's comfort and ease. Many such emotions are now being realized relatively easily. For example, an important emotion that individual humans have had is towards other species. Indeed, an increasing number of people now relate to the well-being of the environment, planet and animals. Individuals associate hope, joy, concern and responsibility for future generations even when they worry about the adverse changes in climate, endangered species and worsening natural ecosystems. Digital transformation offers the ray of hope, amidst concerns. For instance, the World Wildlife Fund (WWF) uses a digital transformation to catalyse its efforts to protect the environment and wildlife, building digital capabilities that aid in monitoring and conservation efforts. However, one key question remains: What is the existential purpose of digital transformation *for humans*?

What are the higher-end thinking and goals that drive humans to digitally transform? Refer to the lion and deer example in Chapter 11. It is only the human being that is thinking about the natural lifecycle, the emotions of the lion and the deer, and has the gun and a vehicle to not worry about being a part of the natural food chain (though that was not historically always true). Humans seem to have outgrown themselves above the natural food chain and like to reflect on the bigger aspects of life (as you are currently doing, thinking about digital transformation and existential purpose). So, what is the uniquely human existential purpose driving digital transformation? Before I answer this question in Chapter 14, I would like to revisit and underline the existential force that underlines the operational purpose of digital transformation in the next chapter.

13

Operational Purpose Revisited

The Force Underlying DaWoGoMo© + Culture

'An organization's ability to learn and translate that learning into action rapidly, is the ultimate competitive advantage.'

—Jack Welch,
former CEO, General Electric

How does a DaWoGoMo© transformation become successful? Each of the transformations (that of digital architecture, work, governance or business model) discussed in Part 3 may lead the organization in conflicting directions. For example, an organization may champion two conflicting transformations: a) enhancing customer-centricity for transforming digital architectures, while b) substituting human workers with technology or reducing their decision rights. These seemingly contradictory moves could have an adverse influence on the organization's culture, leading to resistance towards the transformation process. Getting a singular focus across transformations is the core of success. Scientifically, this requires championing of a singular force for operationalizing the digital transformation. Indeed, a scientific force is latent and hidden (see Chapter 2). Think of gravity. It is ONE force that underlines a lot of phenomena and pervades most physical phenomena on Earth and beyond. I argue that purpose is an important force for operationalizing digital transformation. How does the force of operational purpose manifest?

Digital transformation is a phenomenon that rattles organizations' operations. The operational purpose manifests as a latent scientific force to aid evolution—the existential purpose. Operationally, firms use the extended DaWoGoMo© model. However, success manifests only when their operational goal is the cognitive schema's evolution. As I will explain below, cognitive schema and their collectives (say, teams, functions, departments, business units or the entire organization) evolve in a very similar way: by learning. And learning requires exploration and exploitation. Operationally, firms succeed when they use the DaWoGoMo© model to catalyse learning processes for organizations or individuals (cognitive schema).

Evolution and Individual Learning

How does one evolve in life to become wealthy (like a *crorepati* or millionaire), an astronaut, a marathoner, an ace cricketer or footballer, the top-most civil servant or a leading entrepreneur? Many want to evolve into a better version of themselves to achieve these feats. However, it is challenging to do so. That is, even after one knows *what* she or he wants in life (become a doctor, millionaire and so on), *how* to achieve it is the hard part. How do we evolve in life? The goals are achieved by cognitive schemas that learn. No one knows how to achieve their goals when they begin. Identifying the methods entails exploration and exploitation. Many of you may relate to it. You explore and decide ways to become a sportsperson, engineer or manager and then you exploit regular processes to achieve these goals thereafter.

Formally, exploration is related to the idea of discovery, research, reflection or thinking. Exploitation is very focused and represents action, conduct, enactment or praxis. By nature, exploratory processes are long-term and have unclear outcomes. Basic science is an example of exploratory activity. Generally, exploration is a specialized activity that is distinct from and the opposite of exploitation, as exploitation involves acting upon known principles. Think about how a child takes steps and learns how to walk. Children explore and exploit in order to walk; they explore the world around them to check if they may hold the leg of a particular bed or table and they explore the way their feet land on the ground. They may then find their balance as they put their foot down. As they explore and exploit, they figure out how to take steps and learn the process of walking. Once they figure out the best way forward, they make progress and even begin to run within the next few months or years. This involves repeating the act of walking. So, the first step is exploring how to put their foot right, to learn to take single steps, and after they've done it, they can even espouse to become marathon runners and sprint champions over the next few years.

These central processes for exploring and exploiting continue to guide the evolution. When a grown-up thinks about running a marathon, they just don't get up and run the entire marathon distance of 42.2 km. To run a long distance, one has to explore a lot of different aspects related to physiology, anatomy and mechanics. One figures out the kind of muscle mass or strength required. What is the appropriate food and hydration required? What is the ideal recovery and sleep post-long run? In addition, one exploits the knowledge gained through these explorations by creating and sticking to food schedules, training plans, sleeping and wake-up timetables, managing hydration routines, etc. Beyond running, exploration and exploitation are evolutionary forces core to life. To learn, one first searches for the right things to act upon (explore) and then enacts them (exploit). That is, high-performing individuals engage in the two interlinked activities of exploration and exploitation. The ones who excel master these activities and strike a better balance between the two.

Not surprisingly, the two activities have been historically emphasized across human societies, even from the ancient days. Most societies have had the concept of sadhus (holy men), monks, priests and wanderers who explored different aspects of life (and still do). Many built on their explorations to create routines and exploit their findings. Even today, university systems around the world enable exploration and exploitation by young minds. In universities, researchers develop knowledge and teach it to students, who implement it for their future employers. When done well, the impacts of exploration and exploitation are immense. These lead to breakthrough innovations and organizations. For example, Tesla was able to lay a strong foundation in the electric car industry as it explored the basics of battery design. One of the world's most valued electric car companies was founded with the exploration of the question: Why are electric car batteries not good enough? Indeed, life and organizations evolve by balancing two broad categories of activities: exploration and exploitation. And a successful operational purpose of digital transformation enacts an extended DaWoGoMo© model-based transformation that catalyses these activities.

The operational purpose manifests itself as a force when the digital transformation aids life's (cognitive schema's) evolution. In other words, when transforming using the DaWoGoMo© model, the organization's operational purpose is to catalyse exploration and exploitation. An apt digital transformation may reduce the costs of the two (or make them more effective) for an individual. In the past, individual explorations made limited use of digital technologies. Hence, they were costly. For example, think about the explorations by Christopher Columbus.[1] As far back as the fifteenth century, Christopher Columbus wanted to find India, the preferred trading partner for many Europeans. They wanted to trade for valuable spices, such as pepper, and other materials that may help preserve meat, create medicines, perfumes, etc. Because the Portuguese had established a sea route to India at that time through the southern tip of Africa, Europe wanted to find another route to India. Therefore, Columbus's exploration, enabled by Spanish monarchs, involved a very difficult journey. He did not succeed, and instead of reaching India, Columbus ended up on the shores of the Caribbean. He had found America. Columbus's journey was long, arduous and unknown. More importantly, it was undertaken without the use of much digital technology (an understatement, I realize). Even today, people still have a strong desire to explore. Many of you would go to different cities, states and countries to travel or find business opportunities.

However, unlike Columbus's expedition, which used relatively little technology, today's explorations heavily rely on digital technologies. Subsequently, the costs of exploration have fallen exponentially. Think about the physical and time costs of Columbus's exploration. Comparatively, many people explore the farthest corners of the earth today, taking a comfortable five-to-ten-day vacation. These trips are possible due to the myriad of technologies enabling them. Consider one such technology—Google Maps, which has enabled us to explore foreign lands cheaply—financially and temporally. It has become easier to visit new places in a different state or country. Even a first-time visitor can navigate their way around. Specifically, one may easily explore ways to get from one place to another, manage routes, plan meals and eateries, identify places to tour and so on in a foreign land. As a technology, mapping applications like Google Maps enable exploration and exploitation directly by their users. At large, these aid us in evolving in life.

The Individual vs Organizational Endeavours

Organizations enable individuals to explore and exploit. For example, in the music industry, organizations such as the global player Spotify or the Indian JioSaavn and Gaana allow individuals to explore various forms of entertainment. Specifically, they provide recommendations for music to listen to at a specific time. Recommendations suggest new songs and even new artists. That is, discovery (exploration) is digitized. Such a discovery was cumbersome and time-consuming. Previously, people relied on a large *physical* social network (of like-minded friends). Digital transformation enables these recommendations in real-time. Algorithms provide recommendations using previous data: the listening history of the person or their preferences (for example, what type of music they like to listen to on nights over the weekends), the preferences of other matching profiles and so on. These recommendation algorithms enable discovery that was previously facilitated by friends or family. Similarly, for job seekers, organizations, such as LinkedIn or naukri.com in India, have enabled the discovery of what an individual would like to do, where they would like to work, where their skills may be most suitable or required and so on. Exploring these things in the non-digital age was time-consuming, if at all feasible.

Similarly, much digital transformation enhances individual evolution by enabling its *exploitation*. An example is when an individual connects with friends and family to share news of the birth of a child or a career advancement. Informing a large circle of friends or family is challenging. While one may know all friends and family members, communicating with them all is not trivial. Before the advent of the Internet (just about thirty years ago, for most practical purposes), one had to physically connect, using a phone or face-to-face by travelling. The digital transformation of our networks, via Facebook, Instagram or WhatsApp, has made it easy to connect. It is simple to inform people about what is happening in our lives. Often, people connect in real time, instead of using postcards with pictures or paper-based documents (e.g., e-invites). The technology networks instantly enable us to reach hundreds (or even thousands) of our friends and acquaintances. This would have taken many days to accomplish without digital transformation.

Since exploitation involves executing routine (or well-known) practices, digital transformation may help carry them out faster, quicker and with less effort. Consider the case of online websites enabling house seekers to buy or rent a house. Zillow.com, 99acres.com and MagicBricks.com are some examples. These may also help users assess property valuations. Some have built-in calculators for valuation. These calculators take into account different factors, including: Where is the property located? How old is it? What type of property is it? What size is it? How many rooms does it have? Is it closer to a school? Is it closer to a market, mall or park? Factors such as these are used to determine the price of the house. More accurate valuations may be provided using historical data, the prices of similar properties in that particular area or neighbourhood or whether the property values in that area have changed over time. These help the customer in determining the price directly and quickly. This is especially beneficial, as people may now carry out calculations for many different properties rather quickly.

Many organizations enable balance and combine explorations and exploitations for multiple individuals interacting with them simultaneously. The digital transformation of the mobility (taxi or ride-sharing) industry is an example. Beyond getting from point A to point B, mobility entails customers experiencing a smooth, convenient, hassle-free, clean, efficient and safe ride. Many providers (drivers) who own a car can provide such a ride. However, how do they determine who needs the ride? Similarly, how do customers even know about the best car to take a ride in? These are the questions answered by many ride-sharing companies, such as Ola, Uber and Lyft. These companies identify explorations and exploitations required by both the rider and driver. Further, they recognize that these are exorbitantly complex and expensive endeavours. However, digital transformation offers a solution. The digital transformation of ride-sharing companies frames this as a demand-supply matching problem. They enable the rider to figure out who is willing to offer a ride and help them predict the ride experience with the driver. Similarly, a driver can figure out who is interested in their services at a point in time. The operational force manifests as digital transformation (through the extended DaWoGoMo© model) that enables individuals' evolution—their explorations and exploitations (see Figure 10). And these are being catalysed through *organizations*.

The Organizational Endeavour

Beyond the individual doing so, a collective of individuals (organizations) also pursue explorations and exploitations. A digital transformation with a focus on balancing exploration and exploitation is required for organizational success. Academically, the idea is well entrenched. In his well-known paper, James G. March underlines that the exploration of new possibilities and the exploitation of old certainties underline how an organization learns.[2] In our previously published research, my colleagues and I found that performance is greater in hospitals with digital architectures that enhance exploration and exploitation.[3] The two—exploration and exploitation—influence how human organizations evolve. For example, the ride-sharing organizations (discussed above) explore and exploit them for internal operations. Uber uses digital technologies to explore what customers want, how much they are willing to pay for a new service, when drivers like to operate more and when they don't, when the demand is likely to be higher than normal and so on. Algorithms and related data offer enhanced insights, enabling these pursuits. In one instance, Uber discovered (an unknown) bias created by their algorithms to pay drivers, as it was penalizing women riders for high-quality driving (which took greater time and reduced the number of trips and payouts), amongst other factors.[4] Now consider the traditional taxis, who may never even have the data to explore and find such distortions. Silicon Valley companies, at large, use digital transformation to engage in widespread explorations and exploitations, creating the best possible innovations.

DaWoGoMo© Model to Catalyse Organizational Exploration and Exploitation

Often, individuals explore and exploit within organizations. In an organization, collectively, people come together and interact through digital architectures, work systems, governance mechanisms, business model logic and cultural bonds. Therefore, the operational purpose of digital transformation is to enhance collective explorations and exploitations, or the balance between the two. For example, consider the role of advanced AI. Advanced digital architectures incorporate these technologies for transforming scientific domains like genetics or biology. Using these, scientist teams have been able to analyse large amounts of data for gene sequencing and identify new drug targets. Most notably, consider the speed at which the Covid-19 vaccine was developed during the pandemic. Scientist teams could explore various options. Further, they could exploit the knowledge to come up with a vaccine and save the world.

In general, large-scale data analytics is enabling the discovery of drugs and therapies. Large-scale simulations help assess how different drugs interact with different proteins. In turn, this helps identify treatments. Similarly, in the domain of space exploration, advanced digital architecture transformations are enabling greater exploration and exploitation. The Chandrayaan (or Mangalyaan) programmes by the Indian Space Research Organization (ISRO) underline the tremendous increase in scientific explorations. Advanced explorations enabled the discovery of water molecules on the moon. The Mangalyaan mission placed a spacecraft in orbit around Mars. This is enabling scientific teams to explore Mars' atmosphere, weather and geology. Similarly, scientist teams are using underwater drones to explore oceans as they collect data and images about marine life from the deep sea. In summary, the operational purpose gets its force when the organization transforms the digital architecture to drive exploration and exploitation.

Similarly, the DaWoGoMo© model underlines the role of work transformations. Transforming work may create uncomfortable situations within the organization, as individual pursuits and organizational goals may conflict. Therefore, when organizations focus on exploration and exploitation, work transformation is more focused and purposeful. For example, an environmental organization may use satellite imagery to build new work routines that track deforestation. The adverse impacts on the climate of such deforestation are now being widely recognized. However, quickly assessing any deforestation or loss of biodiversity is hard. This is because large-scale explorations are challenging to carry out manually. Using satellite images, new work routines help explore large-scale areas affected by deforestation. Similarly, work is being transformed across various manufacturing and production operations to enhance sustainability. For example, the advent of 3D printing is transforming manufacturing and may potentially automate it. Producing a large number of complex parts reduces waste. Similarly, in logistic operations, organizations are digitally transforming processes to track inventories. Analytics is helping track shipments, optimize delivery routes and reduce the number of shipments. In turn, these lessen the adverse effects on the climate. In summary, work transformations succeed when they are intended to enhance exploration and exploitation.

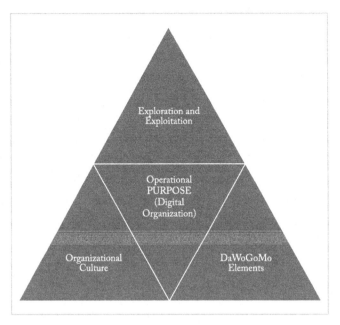

Figure 10: Leveraging the Operational Purpose through the DaWoGoMo© Model

Governance transformation is the third aspect of the DaWoGoMo© model. Such a transformation is effective when it drives individual evolution. The mining industry is one domain where the effects are most obvious. Often, individual workers engaging in mineral extraction have to navigate dangerous contours. This makes mining dangerous. Digital technologies facilitate decisions that require exploration and exploitation. Sensor-based digital transformation helps warn (or pre-warn) if there are places where humans shouldn't go. Decision-making is digitally transformed. This removes the burden on the ill-informed worker to make the potentially fatal decision of entering dangerous underground mines. Similarly, drones and satellites may be used to survey and identify new mineral deposits. This reduces the costly decisions of mining by avoiding failed mining attempts. Governance transformations in agriculture are reducing work and agency costs associated with farming decisions. Farmers use digital technology for precision farming. This enables them to explore the soil and weather conditions across the large farmlands, and explore when it is a good time to sow the seeds or carry out other farming processes, such as irrigation and fertilization. Decision-making costs are reduced as farmers optimize the entire production process for farm produce and agricultural products. The decision rights reduce costs as they enhance exploration and exploitation.

At large, an organization reduces governance costs when it achieves greater exploration and exploitation through the reallocation of decision rights. For example, digital transformation using sensors and analytics helps monitor the performance of equipment and optimize the production rate. Similarly, smart-grid technologies involving the Internet of Things (IoT) are transforming decision-making and governance. These help monitor and manage energy. For instance, organizations using wind farms leverage the power of sensors and analytics to make optimal decisions, maximize the performance of turbines or reduce their maintenance costs. GE formalized the idea through an industrial internet. This helps them: a) explore when a machine is likely to break down or need maintenance and engage in predictive maintenance and b) exploit by optimizing the performance of various machines, such as turbines, and setting the parameters for their operations. GE's industrial internet served as a catalyst for exploration and exploitation, but its implementation required a suitable transformation of decision-making authorities. When such a governance transformation is enabled, the operational force manifests, reducing costs.

Finally, new business models are transforming how exploration and exploitation create value. For example, many travel bloggers document their experiences by using mapping apps to share details of tourist spots, such as caves in Vietnam, coral reefs across the world and other hard-to-find locations. As these travellers and adventurists document their journeys through blogs, they enable exploration and exploitation (through travel) by others. Travel-related explorations were often individual-centric. Today, they aid collective learning. Mapping apps, blogging websites, etc. form the foundations of collective explorations and exploitations. Various advertising-based monetization models create value for these creators. Even in more formal organizations, the transformation of business models to catalyse greater exploration and exploitation creates value. And these are often linked to cultural adaptations.

Consider the Lego case. In September 2004, Lego made an exemplary decision by appointing Jørgen Vig Knudstorp as CEO. It was the first time that a professional from outside the Lego-founding Kristiansen family was at the helm, and a cultural change followed. Lego's dilemma with its business model triggered the change. An iconic name in children's toys, Lego was being challenged by its excellence in the product-centric model and associated culture. It had managed exploration and exploitation very well, as long as the product-centric way was the dominant paradigm in the industry. However, Lego had to adopt a customer-centric model. This was primarily because of the rise of cheaper substitutes, passionate and demanding customers and a wider product variety. Different exploration and exploitation methods were necessary. A cultural and business model change was required to become customer-centric. A slew of organizational changes followed: restructuring of the organization, the foundation of an authentic (as opposed to authoritative) leadership style, reorientation of strategies, community leadership, building new digital technologies and transformation of the relationship with fans and customers. The operational purpose was to enable Lego to explore and exploit differently, with a focus on the customer rather than its product focus. The cultural transformation brought Lego back to success.

Conclusion

Humans (and even animals) are intrinsically motivated to explore and exploit. Organizations are also operationally driven by the same force. If the organization champions exploration and exploitation while employing the DaWoGoMo© model, the force of operational purpose manifests. That is, at its core, the DaWoGoMo© model is operationally effective when it enhances individual and organizational evolution. The learning perspective of evolution underlines greater exploration and exploitation, as well as a proper balance between the two. However, humans are not operationally driven alone. This is because it leaves open the question: *Why* does one want to learn (or explore and exploit)? Indeed, human existential purpose is complex. Deeper thinking requires questioning: Why does digital transformation manifest the way it does? Why do organizations manifest the way they do? What does a human's evolution of life mean? I examine and answer these questions next.

14

The Existential Purpose Computational Freedom

'I think what every entrepreneur in consumer internet is doing is essentially embodying a theory of human nature as individuals and as a group for how they'll react to the service, especially if it's community or network properties, how they'll interact with each other, how this will fit in their landscape of how they identify themselves and how they communicate or transact with *other people.*'

—Reid Hoffman, LinkedIn founder[1]

What is uniquely human about the digital transformation's existential purpose? If we revisit the original example about the chase in the jungle (Chapter 11), only humans are not thinking about survival in the jungle. What is it that has made humans evolve to this level, even when even animals pursue simple emotional goals? Answering the questions requires one to go beyond regular business thinking (as taught in an MBA programme) into philosophy.[2] Digital leaders often incorporate philosophy that offers a lens into life and digital systems. For example, Reid Hoffman, the founder of LinkedIn, best summarizes by suggesting that '. . . philosophy is a study of how to think very clearly.'[3] Specifically, what is the philosophy that guides the creation of large, complex systems in the human world?

What are the unique intellectual and philosophical foundations that guide humans to go above and beyond other species and build complex digital networks of humans, such as the Open Network for Digital Commerce (ONDC), LinkedIn, WhatsApp, Instagram and so on? A good starting point for answering this question is to consider the force that drives our actions and behaviours.

The Neuroscientific Wo(man) and Reward

Why do we act and what guides human behaviours? This has been a well-studied question in human motivation literature. Scholars identify different ways to determine human motivations, such as focusing on needs or individual personalities. Indeed, individuals are found to have different personality characteristics (such as the ones identified by the Big Five personality characteristics model)[4] and these shape their preferences and motivations. To identify the unique human existential purpose, instead of focusing on needs or personalities, I continue to conceptualize an Individual (as an information processing entity) that gets its existence from the state of neural networks (revisit Chapter 11). This neuro-scientific conceptualization of a wo(man) underlines that the purpose motivating humans to act is 'rewards'.

Individuals seek rewards. Various research studies reveal how physiological processing and behavioural responses are driven by the pursuit of rewards.[5] Indeed, reward system activations influence economic activity.[6] Experiments at the SPAN lab (Stanford University) reveal how stimuli (money) activate university students' brain reward management systems. Investors are also known to take risks (even become impulsive) when the reward systems are aroused. Fear of losing (rewards) enhances timidity. Richard L. Peterson argues that these reward-related dynamics are strongly prevalent across the economy and may further lead to stock market bubbles, greater consumer purchasing or credit offtake.[7] The neuroscientific basis for the pursuit of rewards is now well known. The neurotransmitter dopamine is one crucial element that helps sense rewards.[8] Many other parts of the brain are involved, too. For example, using neuroimaging, studies have found that the ventral striatum is activated when anyone is presented with natural rewards such as food (e.g., chocolate)[9] or mate attributes such as beautiful faces.[10] Similarly, the caudate nucleus,[11] limbic system (amygdala, anterior cingulate cortex and nucleus accumbens) and mesocorticolimbic DA systems are involved in sensing rewards.[12] Peterson dissects the reward circuitry, identifying the brain region that '. . . runs from the midbrain through the limbic system and ends in the neocortex'. [13]

Further, reward-related dynamics are complex. For example, insights into reward expectations and reward realizations underlined different neuroscientific dynamics. Similarly, losses and gains in rewards involve different neuroscientific processes. The complex neuroscientific dynamics help us unravel that the *pursuit of reward* is neuroscientifically hardwired in humans and drives the digital transformation's existential purpose. Why? Because it helps us relate digital technologies with bounded rationality.

Bounded Rationality, Digital Transformation and Neuroscientific Reward

Bounded rationality is the reason digital transformation is core to human existence. Two facts help us understand the dynamic: a) the brain processes information to realize rewards and b) it has limited abilities to do so. First, information processing is the neuroscientific brain dynamic that helps individuals realize rewards. This is because computations underlie all cognitive tasks and processes. Neural networks process information in order to work. For example, Todd Braver and colleagues find a linear correlation between task-induced working memory (WM) and activity in the dorsolateral and left inferior regions of the prefrontal cortex (PFC).[14] Similarly, Jonathan Cohen and colleagues demonstrate that working memory influences the performance of higher-order cognitive functions, such as planning and problem-solving.[15] Specifically, they find that distinct brain regions manage the two well-established brain processes—executive control and active maintenance. Notably, they find that the dynamic maintenance process, which aims to assure the availability of information in working memory, involves both the prefrontal cortex and parietal cortex. This was quite contrary to the notion prevalent before 1997 that the prefrontal cortex contributed only to executive control. While interesting findings such as these are emerging and much remains unknown, it is clear that computations underlie human activities.

The fact that humans compute is not an ordinary revelation. It took years to realize this. Indeed, the history of the human brain has a long history of research. Until about AD 2, Egyptians considered the heart to be the source of intelligence.[16] Hippocrates first conceptualized the brain as the key organ that facilitates superior sensations, aspirations and cognition. After René Descartes, the seventeenth-century French philosopher, established that the brain is the organ that computes (linking it with other complex machines like clocks), other scientists—Thomas Willis, Luigi Galvani, Alessandro Volta and Emil du Bois–Reymond—unravelled the dynamics related to neuron structure, fluid flows, electric impulses and cell culture. In the early twentieth century, joint winners of the Nobel Prize for physiology and medicine in 1906, Camillo Golgi and Santiago Ramon y Cajal unravelled the brain's information processing dynamics.[17] The identification of neural networks builds on this history. Over the last century, researchers have established that the presence of a neural network enables humans to process information.

Second, bounded rationality emphasizes that neural networks have limited information processing capabilities, resulting in individuals not making the most optimal choices. Bounded rationality is a concept that originated in the writings of Herbert A. Simon.[18] Because of bounded rationality, humans use decision models that are appropriate or good enough but may be *irrational* (illogical or suboptimal).[19] Indeed, satisficing (or good enough) decisions may be used (and may be sufficient) for various tasks, such as reading. For example, individuals may understand a sentence only by selectively reading some words and skipping the rest, even making perfect sense of an entire sentence of gobbled-up words. While it works well for reading, satisficing falls short in many cases, suppressing rewards. For example, while grocery shopping, one may choose a fruit (say, a banana) based on the 95 per cent accuracy of hue perception (and probably access to its smell), even when a sure way to avoid indigestion would require an assessment of its precise spectral reflectance.[20] Emphasizing the role of bounded rationality for choice-making, James March summarizes the trade-offs: '. . . bounded rationality has come to be recognized widely, though not universally, both as an accurate portrayal of much choice behaviour and as a normatively sensible adjustment to the costs and character of information gathering and processing by human.'[21]

In general, bounded rationality underlines the inherent superior ability of digital technologies over that of individuals. For instance, while human minds excel at recognizing patterns, even the best experts can identify only up to 1,00,000 patterns. In 1997, noted chess champion Garry Kasparov was able to memorize 1,00,000 board positions; Shakespeare could analyse 1,00,000-word senses; and medical specialists may recall up to 1,00,000 concepts.[22] Digital technologies are far superior to human pattern recognition. Games like chess most clearly demonstrate this. For example, Deep Blue defeated the chess champion Kasparov. Many contemporary computers exceed Kasparov's abilities for analysing board positions. Neuroscientific constraints lay the foundations of bounded rationality. James R. Bettman underlines that memory storage (often done in chunks) has limited capacity.[23] Fixing one chunk of memory in the long term takes about ten seconds, while recalling the chunk later takes about five seconds.[24] Similarly, the human eye contains approximately 130 million photoreceptors responsible for creating visual snapshots, each of which may create approximately 15 MB (megabytes) of data. At large, humans are limited in their ability to process information (say, visual), as many insects can *see* wavelengths in the electromagnetic spectrum below 400–700 nanometres—the range discerned by humans.

Similarly, neural processing takes approximately 10 ms per synaptic stage, and this is much less than the speed at which modern computers process information.

Bounded Rationality and Life

The adverse effects of bounded rationality on rewards are evident in their impacts on life outcomes, as people may fail to plan or think correctly. Many brain processes are largely controlled by the prefrontal cortex (PFC)—the executive area that integrates information from other brain regions.[25] The PFC helps create plans and goals. Computations underline such brain processes. However, because of limited computing capabilities, many individuals fail to execute these business processes. Automated brain processes are one example of such failure. These are associated with a lack of conscious decision-making. That is, automatic processes are evolutionarily wired and may be triggered with little or no deliberation.[26] Brain injuries or any distortions (due to stress or an imbalance in neurotransmitters) may trigger these processes as well. Generally, these processes are a representation of limited computational capabilities and their adverse outcomes manifest.

An example is what happened (and usually happens) when one person used social media without much thought. On her way to Cape Town (South Africa) from the John F. Kennedy International Airport (New York), senior director of corporate communication at IAC, Justine Sacco, tweeted, 'Going to Africa. Hope I don't get AIDS. Just kidding.

I'm white!' The tweet went viral and she was soon fired.[27] Limited computational abilities led her to say those ill-fated words. A more thoughtful and well-computed tweet may have had the opposite effect. In general, individuals may lack the capability to compute at certain times. And, because of limited computational capabilities, brain processes may lead to bad judgements and social ills or biases may set in. For example, one may automatically consider a face 'horrible' because one is conditioned to do so.[28] Many more problems manifest because of the underlying lack of computational capabilities in an individual. A student may select the first college they come across that seems like a good fit without researching other options; an investor may overlook important information—while investing—about a company's financial performance; a company may choose to cut costs while ignoring the effects on customer loyalty.

Digital Transformation and Externalized Computations

Not surprisingly, the information processing done by the brain is more commonly being enhanced through various artificially intelligent technologies within organizations. Robots are processing information that may help conduct brain surgeries. Similarly, patient-centred systems and databases are enabling patients to make choices about their health.[29] Further, autonomous vehicles can engage in complex information processing, instead of human drivers.[30] Throughout existence, human beings and other living species have been the only entities able to compute. This is because information processing has been exclusively restricted to brain regions for most of human evolution. The prevalence of AI, or intelligence that is not biological, is rather new.[31] That is, externalization of computations is a recent phenomenon.

Externalized computation is the idea that I talk about to discuss the similarities in the way human beings' neuroscientific processes are imitated by digital technologies that compute. Externalized computation enabled by digital technologies is disrupting the tight coupling between computations and living beings. For example, let us revisit our example of the chase in the jungle (see Chapter 11). The well-being of a living being—either deer or lion—depends on the superiority of their computations. In the chase in the wild, the deer computes how far the lion is and the lion computes which way the deer will go, as each decides upon the course of their run. The superiority of computations, regarding their paths, strategies and the forest landscape, determines who survives. In the early ages, computations also determined the outcome of competition between animals and human beings. For example, during a hunt, human beings could compute the location of a deer (taking into account the speed of their arrow and that of the running deer) to get the hunt. Similarly, the deer would use its computations to decide on a certain location to run into. Based on the superiority of computations, either the human got the hunt or the deer got its life.

In general, the computations were performed by living beings only (and not by, say, the arrow). Today, modern missiles can compute. That is, the externalization of computations enables technologies (such as a missile) to process information, changing the balance of power on the planet. Compared to physically stronger animals, human beings lack physical speed and dexterity and have created a complex landscape of computational machines. These machines are enabling externalization of computations in place of biological neurons, such as through missiles that can identify the target and hit it, sensors that can identify if a person is passing by and switch on the lights accordingly, or ChatGPT that can automatically create a text for a speech or the code for a complex algorithm. In general, externalization of computations means making machines (such as calculators and computers) an alternate way to compute, beyond the cognition within living beings. Why are external computations mushrooming? Because humans do not like to compute everything and seek *computational freedom*.

Human Existential Purpose: Computational Freedom

Computational freedom is the reason (the why) that motivates individuals to transform digitally by harnessing externalized computations. That is, successful digital transformation's existential purpose is almost always to enhance computational freedoms. For example, the advent of Covid-19 severely restricted the freedom to move around for a vast majority of the population across the globe and digital technologies freed people. The technologies helped overcome the suppressive forces. For example, information technologies helped tackle the spread of Covid-19. Across the world—in the US, India, Taiwan, China and Singapore—leaders relied on digital platforms to strategize their fight against the virus. Most countries rely on different technologies and algorithms to monitor and manage cases. Similarly, technology allowed people to communicate and work even when the pandemic limited their freedom to interact in a face-to-face mode. In other words, the pandemic demonstrated how bounded rationality manifests. Generally, beyond crises, the digital age is bringing freedom to the masses. In her address to the Massachusetts Institute of Technology's (MIT's) Class of 2018, Sheryl Sandberg, former Facebook COO, summarized it this way:

> Today, anyone with an internet connection can inspire millions with a single sentence or a single image. That gives extraordinary power to those who use it to do good—to march for equality; to reignite the movement against sexual harassment; to rally around the things they care about and the people they want to be there for.
>
> But it also empowers those who would seek to do harm. When everyone has a voice, some raise their voices in hatred. When everyone can share, some share

lies. And when everyone can organize, some organize against the things we value the most.[32]

Living beings generally acquire many digital (and other) capabilities existentially, aiming to enhance their freedom—the ability to compute when and what they desire. That is, digital transformation is driven by the *pursuit of computational freedom*.

While bounded rationality makes one seek externalization of computations (and computational freedom), the crucial question is: Why? Why do humans seek computational freedom? Because it helps them evolve. Digital transformation frees the human mind to evolve by taking over computations.[33] Imagine yourself taking a traditional taxi, which many of you may have taken at some point in time. Compare the experience with that of using an Ola (or an Uber) cab. Is the freedom you sense different? An Ola (or Uber) cab aids your ability to think about the expected drive—the quality of the car, the driver and so on. You can read the reviews and driver ratings that you did not have access to in the case of a traditional car service. Similarly, your ability to see the pricing (and not have to negotiate the same) is enhanced. Think about the value of a non-interactive price determination, as is the case with ride-sharing, and think how valuable it is when you are travelling in a foreign country where you do not understand the language. Ride-sharing enables you to take a ride without even the need to talk to the driver—a tremendous relief if the communication is not feasible or desirable. The computational freedom for both passengers and drivers is immense—the former gets to travel without knowing or discussing details and the latter gets the business where none may have been plausible otherwise.

Conclusion

Digital technologies and transformation are driven by the unique recognition of computations. At their core, humans compute information. That is, *information* is a distinct entity that can be computed. Information may comprise the REM sleep cycles of a person, the temperature of a steel furnace processing metals to make steel and information collected from a cow's abdomen about a sickness it may be carrying. Moreover, the digital age is characterized by advancements in the storage and processing of this information (revisit Chapter 1). For example, various digital hardware—sensors, keyboards and other electronic devices such as cameras—are being used to collect and create information. Electronics, computers and information systems are some of the sciences that help develop principles to study the computations of information. For example, great advances have been made in a) statistical, econometric and machine learning methods and b) the availability of electronic hardware (e.g., Hadoop, in-memory computing, supercomputers, PCs and cell phones). These advances are leading to digital technologies that are remarkably different from others in previous ages, as they *externalize computations*.

Technologies in the previous era used electrical, mechanical or chemical concepts. They had limited abilities to compute, compared to the digital technologies of today. Even the first generations of digital technologies (e.g., databases) were much simpler. Over the last few years, developments in electronics and computational sciences have ushered in advanced new-age digital technologies that are transforming the locus of computations, externalizing them rapidly. Modern technologies, such as those using machine learning or deep learning—broadly defined as artificially intelligent technologies—can overcome a lot of constraints on human information processing. Successful digital transformation leverages the potential for externalizing computations to change and further the human desire for computational freedom. Because the ability to externalize computations that advance biological capacities is going to only increase, the urge to enhance computational freedoms will increase. Computational freedom has a strong neuroscientific basis. Simple creatures may have a few hundred neurons, and the human brain may include billions of neurons. It is for this reason that the human mind can and wants to unravel complex philosophies that aid the evolution of life. Computational freedom aids in these endeavours. As argued by V.S. Ramachandran, a leader in the domain of cognitive neuroscience:

> . . . human brain . . . this mass of jelly, three-pound mass of jelly you can hold in the palm of your hand, and it can contemplate the vastness of interstellar space. It can contemplate the meaning of infinity and it can contemplate itself contemplating on the meaning of infinity.[34]

Computational freedom enhances the human computational ability to reflect on the purpose of self, computations and existence. A lot of history underlines the power of freedom to drive humans.

When the forces suppressing individual freedoms are continually present, it hinders life's growth. Human beings consider it rewarding to overpower these forces. The great Indian poet and the winner of the 1913 Nobel Prize in literature, Rabindranath Tagore, wrote a well-known poem expressing his wish for Indian freedom from the British. Beyond foreign occupation, Tagore wished his motherland, India, was free from fear and illogical and orthodox beliefs and traditions. Indeed, restrictions on freedom may arise due to sickness, diseases, poverty, genetics and other factors. These force individuals to compute in ways that they do not wish to. Unwanted computations create a negative outcome—one that may hinder emotional evolution. Therefore, individuals seek computational freedom. In other words, digital transformation is driven by an existential purpose that liberates humans from thinking about things they do not desire to deliberate upon.

While computational freedom manifests at the neuroscientific (the individual) level, how does it define the existential purpose of the organization? That is, what is the link between individual purpose (enhancement of computational freedom) and the creation of a digital organization? I present and discuss this next.

15

Digital Organization

'We are in one of the most remarkable moments in human history and you will not just live through it, you will shape it. Many of you will work on technologies that will change the world. You will connect the rest of the world, create new jobs and disrupt old ones, give machines new powers to think and give us the means to communicate in ways we haven't even thought of.'

—Sheryl Sandberg, former COO, Facebook, in her address to the MIT Class of 2018[1]

'AIs will dramatically accelerate the rate of medical breakthroughs. The amount of data in biology is very large and it's hard for humans to keep track of all the ways that complex biological systems work. There is already software that can look at this data, infer what the pathways are, search for targets on pathogens and design drugs accordingly. Some companies are working on cancer drugs that were developed this way.

'The next generation of tools will be much more efficient, and they'll be able to predict side effects and figure out dosing levels.'

—Bill Gates, founder of Microsoft[2]

Individuals are hardwired to seek computational freedom, but why is an organization required to achieve it? It is important to answer this question to understand the existential purpose of *organizational* digital transformation. What drives an organization's existence should also drive its digital transformation. When the industrial age started, the production function provided the intellectual foundations for an organization's transformation (see Chapter 2). The goal of any such transformation was to enhance productivity. However, in modern business and society, creativity, not productivity, has emerged as the driver of superior performance. Understanding the organizational existential purpose helps us creatively champion a successful digital transformation. So, why does an organization exist?

Organization: A Collective of Individuals Seeking Computational Freedoms

Why do individuals organize? The existential purpose of the organization is a very permanent phenomenon on this planet. Organizing goes beyond human beings. Most living beings organize their lives. Birds organize by creating nests; a beaver is known to build dams and many animals store food in the summer. Indeed, humans organize to give their young ones a safe place to sleep and play, as well as satiate their hunger during the winter by consuming food stored in warehouses and growing food in the summer. The emotional goals (outlined in Chapter 11) may be the reason for individuals to organize in the animal kingdom. However, human organization goes beyond these; it is the most complex among all animal species.

The complexity is apparent even when humans organize in families. As an intricate organization, a successful family provides humans with a space to interact and discuss their views, fathom the depths of themselves and humanity and learn how to express their feelings of concern, joy, frustration, anger, pain, ambition or sorrow. Humans also organize to unravel their religious hearts and minds as they visit and congregate at places of worship. Further, humans may organize so they can:

1. Satisfy themselves by climbing a mountain (say, Mt Everest) or venturing to explore nature or space (say, by going to the Moon).
2. Heal a broken bone using complex medical procedures.
3. Satiate their intellectual cravings, say, by understanding how the world works; for example, by proposing and testing theories of gravitational force, string theory, human networks or communications.

Organizations exist because achieving these outcomes is computationally complex and daunting for any one individual. Individuals cannot manage the computational complexity by themselves. So, human beings organize collectively to enhance their computational freedoms. That is, organizations help individuals evolve by freeing them from thinking about things they don't want to think about. When we are freed from being required to compute about A, we get to think about B. For example, when not fed, the individual thinks about food (A) all the time, failing to focus on other pursuits. Such individuals do not get to think about work, algorithms or the universe (Bs). So, individuals organized to create food chains (encompassing agriculture, fertilizer factories and other related entities) and inventory them (through the retail industry, for example). Specialized organizations—managed through a complex economic system—made it feasible for humans to stop thinking about food so they could focus on spending time with their children, learning a language or skill, watching an entertainment show, preparing a rangoli for a festival, running experiments, creating technologies and so on. That is, collectively, the organization of individuals conserves the individual computational burden. How does digital transformation enhance human endeavours to organize?

Digital Transformation and Artificially Intelligent Ecosystems

Digital transformation requires us to rethink organization. An organization creates artificially intelligent ecosystems (AIEs) that are required to overcome bounded rationality and enhance computational freedoms. As the word indicates, an AIE is a complex intertwined system comprising products, services, teams, norms, regulations, laws and other such elements. Beyond technology, an AIE underlines the force of the human collective. The organization manifests as it gives a place for people (employees, customers and other stakeholders) to think, reflect, learn, discuss, debate and act. Many organizations may interact to aid in the creation of AIEs. Intelligence is distributed across society, the press, educational institutions, nature, the universe, regulatory and legal bodies, employees, shareholders, customers and others. The existential purpose of *any* organization is to build AIEs that enhance computational freedoms by leveraging this intelligence through technologies. Modern AIEs are complex, with digital technologies at their core. However, even non-digital technologies are intertwined with AIEs.

Calendars are an example. A non-digital technology, the calendar is associated with deep knowledge of astronomy, geometry, philosophy of time and mathematics. Further, it enabled creating an intelligent ecosystem of activities around itself. The earliest calendars were based on lunar or solar cycles (used by ancient civilizations such as the Sumerians, Indians, Egyptians and Babylonians) and helped determine the timing of planting and harvesting crops, religious festivals and other activities. These helped organize time, freeing humans from a lot of different computations that would be required in a non-calendar world. Calendars unleashed computational freedoms as they enabled us to plan, organize and coordinate our activities, from agriculture and trade to social and religious events. Much of the disorder in the human world is managed through culture, society, religions or faiths, and this requires synchronized coordination enabled through an ecosystem built around a calendar.

Today, digital transformation is enabling the development of much-advanced AIEs that exponentially overcome an individual's computational constraints. I discuss two AIEs that are digitally enabled: smart homes and autonomous cars.

a) Smart Homes: AIE, Case # 1

Since the earlier days, a big computational challenge for humans has been related to their homes. There was a time when human beings lived in caves. However, this was a computational challenge. Imagine a family sleeping through the night in a cave. Parents have to wonder all night if there may be a predator that may endanger their families, especially young children. All night, the brain computes, listening for any sounds or other indications of a threat. It is similar to being at war all night. How does the house resolve the computational challenge? Think about a basic house with a proper door lock. This house is an example of an AIE. It ensures that no animal or another unwanted human can walk in with bad intentions. This simply intelligent ecosystem provided by a home saves so many computational challenges for an individual, as one does not have to be on the lookout 24x7 for safety. Modern houses are much different from the early simple houses, with smart devices and equipment that control appliances, monitor rooms and external doors, control temperature, amongst other uses. The computational complexities of many types are now resolved through these smart homes, as they:

- Use data and analytics to optimize energy usage and improve efficiency.
- Apply smart home technologies to build cyber-physical systems accessible over mobile or the Internet.
- Integrate smart home devices, such as smart thermostats, speakers, lighting systems, refrigerators, locks and security systems.

Smart homes enhance our sense of security and comfort, freeing up our time for other activities. That is, they enhance our computational freedom. In the US, for instance, automated alarms link directly to emergency services (like 911). This ensures a prompt police response to break-ins or other security threats. Similarly, smart homes eliminate the need to constantly think about nightly routines. Instead of manually turning off lights every night at 9 p.m., you can simply set an automated routine. This saves you time and mental effort spent remembering and walking around to turn them off. Systems like Nest or Google Home also save you the trouble of monitoring the temperature inside the house or manually maintaining the desired temperature. These modern smart thermostats monitor the temperature automatically, ensuring your comfort without any extra effort on your part.

b) Mobility and Autonomous Car: AIE, Case # 2

Throughout history, humans have faced barriers to moving around. However, the wheel-based cart generated an AIE that helped them transport products over long distances. The wheel was an intelligent innovation, as it revolutionized transportation by harnessing the role of friction. The wheel-based cart was an example of an AIE of the time, as it freed people, giving them the superior capability to carry goods. People had to think less about moving goods over small distances. The evolution in the ecosystem over time led to steam-based wagons and then modern cars, making people think less and less about travelling to even far-off distances. Think about the computational burden when one had to travel even 500 miles/km, say, 1000 years ago, versus doing so today. The evolution of vehicles (think, cars) has transformed what travel means. The older cars just took you from point A to point B. Today, the introduction of DVD players, music, lane assists, automatic brakes and so on makes travel a more fun and safer experience.

Digital transformation is making mobility more intelligent. The driving ecosystem was catalysed by the automobile car and intelligent systems evolved with the growth of highways, traffic regulations, traffic police and so on. Digital transformation is leading to the advent of autonomous cars, which are creating an even more advanced AIE. Google Car is a case in point. Google is a technology giant with exceptional abilities to create computational capabilities. Sebastian Thrun, former Google VP, and fellow and professor of computer science at Stanford University, participated in a DARPA challenge to create autonomous cars, primarily because he was motivated by his desire to do something to save the accident deaths (due to a personal tragedy).[3] The linking of Google with the mobility-enabling ecosystem (the car) underlines how the car (and driving at large) is a computational exercise.

At its core, driving is a computational process. Drivers constantly capture and process information—the road layout, vehicle speeds, surrounding objects and people—to make decisions about acceleration, braking, turning and other manoeuvres. Traffic conditions can significantly increase the complexity of these computations.[4] Imagine the multitude of people making these calculations on the road, their decisions influencing one another. When you drive at high speeds, your safety depends not only on your own skills but also on the computational capabilities of those around you, including new or inexperienced drivers. The constant influx of new drivers into the system highlights the computational dependency of driving. Our safety hinges on the abilities of others, many of whom may be just learning.

Consider the potential benefits of all drivers having the equivalent of six months to a year of experience when they start—a significant improvement in overall safety.[5] Autonomous vehicles, with their superior computational capabilities, can process information from even greater distances, creating a safer driving environment. Modern features, such as lane assist, are early examples of these advancements. Self-driving cars have exponentially greater computational capabilities, enhancing computational freedom, particularly for the blind or visually impaired. Even for sighted individuals, these vehicles can be a boon, reducing the cognitive load associated with navigating traffic, especially when they are returning home after a long day at work.

The autonomous car's proposition for computational freedom requires deep insight. Consider whether we know how to operate the machinery for the numerous products or services that we consume. Just as it's not necessary to know how to operate a flour-milling machine to enjoy roti or bread, we don't need to be experts in driving to enjoy the benefits of mobility. The idea has been brought to the fore by various modern taxi companies, such as Uber, Lyft and others. An AIE will need to evolve around the self-driving car in order to realize the benefits of its intelligence, due to its superior computational capabilities. Creating such an AIE requires all stakeholders to see the value in the proposition:

1. Customers find it valuable not to drive:
   a. They can move around and spend time with family while driving.
   b. They have much reduced chances of being in an accident.
2. Employees see value in creating such a car:
   a. Sebastian Thrun is driven by a personal focus.
3. Investors see the value propositions in engaging:
   a. Google thinks about Moonshot projects and is willing to invest money in such projects.
   b. The idea fits with the philosophy of digitizing the world's information.
4. Partners find value in the self-driving car:
   a. Google is a member of an open automotive alliance that encourages its partners to develop self-driving cars.
   b. Technology providers, such as Lidar, support the automation of driving.
5. Regulators see value in self-driving cars:
   a. Managing self-driven cars requires the entire traffic ecosystem, e.g., lane markings, red lights and so on.
   b. Regulators need to create laws in case there is an accident. Who will bear the liability—the software developer or another entity?

AIEs hold immense promise, as new technologies like ChatGPT and other large language models (LLMs) are causing disruptions across organizations. As discussed throughout the book, the growing capabilities of AI are fundamentally challenging the way we work and think. AIEs offer a framework for harnessing the potential of AI while mitigating its risks. However, there is the crucial question of whether or not AIEs are essential for our very existence. And, what is the existential purpose of an organization? Though seemingly complex, the existential purpose of the organization becomes crystal clear when seen from a child's perspective.

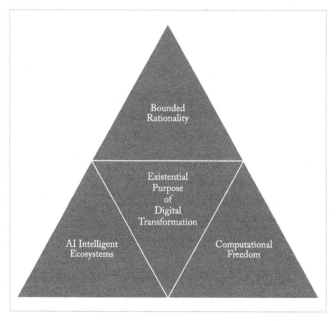

Figure 11: The Existential Purpose of Digital Transformation

Entropy: Why Organize?

I once happened to ask a child: Why do you think we organize our beds? His answer was simple: because we would be disorganized if we didn't. Indeed, things tend to be disorderly. This is what the law of entropy asserts. It comes naturally to children. Children naturally organize their surroundings—sorting toys and building structures— to make sense of the world and to interact with it more effectively. I wish everyone else remembered this! The fundamental existential purpose of organizations is to counter entropy (disorder). Beyond human beings, most living beings organize their lives to avoid disorder. Even animals have the intelligence to know and sense disorder. But why is disorder a concern? Because the disorder is inherent, as outlined in the fundamental law of physics, the second law of thermodynamics. The law outlines entropy as a central property of the universe.

Entropy, a concept studied in various ways, has its origins in thermodynamics, where it reflects the natural tendency towards disorder. The second law of thermodynamics underlines that in isolated systems (closed systems with no exchange of matter or energy with the surroundings), entropy always increases. Building over the works of Said Carnot (1796–1832) and William Thomson, renowned as Lord Kelvin (1824–1907), among others, Rudolf Clausius (1822–1888)[6] proposed a version of the second law of thermodynamics explaining why heat always flows from a source with a high temperature to a source with a lower temperature. In an isolated system, a hot object will cool down, not vice versa.[7] Similarly, a drop of ink in water disperses and sugar dissolves in water because the resulting mixtures have higher entropy (greater disorder) than the initial separated components.[8] This highlights the law's core principle: the universe's overall entropy (disorder) is constantly increasing. Entropy is generally a measure of randomness or disorder in various scientific domains, including management.

Artificially Intelligent Ecosystems and Entropy

AIEs can be powerful tools in combating disorder. Organizations are existentially driven to evolve. Positive emotions enable evolution. Entropy (the tendency towards disorder) acts as a force against it. When faced with entropy, our ability to assess disorder helps us act to counter it. This assessment guides us towards what we perceive as the 'right thing to do'. Often, this requires us to organize, as the right thing is not feasible for an individual to accomplish alone. Digital transformations empower us to evaluate disorder and respond to it through more effective organization. Doing so enhances our computational freedoms and requires building AIEs (see Figure 11).

In essence, successful digital transformation is an organizational effort to assess and counter disorder by making choices that combat entropy. It enhances individual freedom by creating AIEs (see Figure 12). Let's revisit the example of the chase in the jungle (introduced in Chapter 11). In my years of teaching at IIM Ahmedabad, I've discussed this scenario with various audiences—senior executives, company heads, students, managers and more. Often, a majority express a reluctance to interfere with nature's lifecycle. They prefer a hands-off approach, letting nature run its course. Why? Perhaps there's a deep sense of existence that we share with nature and that we're all aware of. Alternatively, as I often find, there is also inconsistency and unclarity about existence.

Our behaviours reveal our existential purpose more clearly. We've undeniably influenced nature, sometimes irresponsibly, by causing significant damage to environments and wildlife habitats. In the widespread natural transformation that surrounds us, trees have been cut to make tables and chairs; you're probably sitting on one right now. Natural resources are used to create clothes, laptops and so on.

In the jungle chase example, there are three distinct species: the lion, the deer and the human. Humans are the ones who can (and do) think about the computational freedoms of all. This is primarily because of our heightened sense of disorder. Human beings are the topmost transformation agents in the known universe because of this sense. The ideas of fairness and equity are regularized in the human world. Humans have implemented complex AIEs, such as the police force and the legal system, including animal rights laws; they have developed the Internet, mobile phones and app ecosystems (Android or iOS); and they have even created pet foods and so on. Indeed, in the example above, it is only the human who is thinking about the larger issues of society and digital technologies—their interference (or not) with nature, animal protection and responsibility to other species, food storage, environmental transformation and climate change. Humans have been and will continue to be the change agents in nature. As highlighted by Sheryl Sandberg, former Facebook COO, in her address to the MIT Class of 2018:

> ... We are not passive observers of these changes. We can't be. Trends do not just happen, they are the result of choices people make. We are not indifferent creators, we have a duty of care and when even with the best of intentions you go astray, as many of us have, you have the responsibility to course correct. We are accountable to the people who use what we build, to our colleagues, to ourselves and to our values.[9]

Digital transformations require one to think about the existential purpose of the individual (computational freedom) and that of the organization (building AIEs) as the two are intertwined. Successful digital transformation manifests itself when technologies are leveraged to enhance the intelligence of artificial ecosystems. The digitization of entertainment and retail underscores this even more.

First, entertainment has been eternal throughout our existence. Today, Netflix has transformed the meaning of entertainment for many. Disorders faced by many can be quantified through the idea of unit economics. Specifically, the digital transformation of entertainment improved the unit economics underlying entertainment.[10] Early on, the concept of *numaish* or *melas* (fairs) brought people out of their homes as they sought to be entertained. Cinema halls formalized entertainment as an outdoor activity, offering people the chance to sit comfortably in a chair and get entertained through the big screen in dark rooms. The disorder was reduced as consumers began to prefer cinema halls over fairs. Viewers felt less disorder and they could appreciate nuances in the content. This led to the evolution of the content. As the customers appreciated nuances in acting, the cinematic actors evolved as well. Further, professionals—lyricists, singers, actors, directors, screenplay writers and others—evolved to become more specialized as nuances in content were demanded. Such nuances in filmmaking may be lost in the din or noise of a fair.

Next, television drew people back to their homes for entertainment, thereby further reducing disorder. With the comfort at home being even greater than in a cinema hall, people could now be entertained in their own context (e.g., wearing whatever they find comfortable). DVD or video rentals still require people to step out—an inconvenience (leading to a heightened disorder) as the time available for entertainment may be limited for many consumers today. Over-the-top (OTT) platforms are streaming entertainment content directly into homes on demand. These save time and reduce computational burden, as they personalize recommendations to help discern the content likely to be most entertaining for the consumer, at that time.

Second, retail is a very good example of how digital transformation has reduced disorder. In early India, consumers largely shopped through haats; some still do. In the past, purchasing a pair of pants or a suit meant waiting for the haat, which only occurred every two to three weeks. Think about the individual effort required. Attending the haat required walking miles to attend the once-in-a-fortnight event. At the haat, they could only choose from a limited set (say, three) of colour options. Still, the whole experience was emotionally rewarding back then. In comparison, consumers today are more demanding. They seek instant gratification. They expect to be able to shop at midnight, within a few seconds of thinking about the product they want to buy. Waiting for a fortnight is not an option. Well, waiting even a day is too long in many cases. Let alone three, they want access to a wide variety of options to choose from. They also expect detailed customer reviews for all the options from consumers of the product being evaluated.[11] While examples are many, the existential purpose underlines that digital transformations requires synchronizing the two: our computational freedom (life's existential purpose) and AIEs (organization's existential purpose).

Conclusion

Digital transformation goes beyond the act of digitizing human organizations. The endeavour gets its strength from the computational freedom it engenders. It is often beyond any individual's capabilities to compute the secrets of the universe, the intricacies of the human genome, ways to develop vaccines and medicines and so on. This inability to compute on their own leads humans to organize. Therefore, digital transformation is intended to aid the collective enterprise of individuals to evolve by computing.[12] This purpose has driven us all to organize our living history. Many organizations have freed individuals to search for the truths of the world, children to go to school and even adults to pursue their passions. So digital transformation entails synching the individual and organization's existential purpose—the two sides of the same coin.

Indeed, successful digital transformation enhances the AIEs significantly. Lives are influenced. However, success in digital transformation manifests when organizations recognize the individual disorder. The influence of a responsible existential purpose on computational freedom is significant. Without a clear sense of disorder, the freedom may decrease and result in human digital organizations perpetuating the ills of the past. Being responsible and careful towards transformation is crucial, and the existential purpose helps the organization do so. Indeed, the transformations championed by humans have not been very reliable, leading to a transformative paradox (see Chapter 1). That is, a responsible fight to save the environment and climate and achieve other great goals, which requires a focus on individual computational freedoms. Because any transformation by humans affects the planet and its inhabitants, we need to clearly think about the purpose—the existential purpose.

When championed with the existential purpose of taking the onus for different species' computational freedom and overcoming their limited computational capabilities, the digital transformation represents the apt fight against entropy. Digital leaders who conceptualize and think about AIEs that achieve these goals are the ones who succeed in their digital transformations the most.

Part 5

Enacting a Digital Transformation

16

The Purpose

> 'Computer software changes how architects think about buildings, surgeons about bodies, and CEOs about businesses. It also changes how teachers think about teaching and how their students think about learning. In all of these cases, the challenge is to deeply understand the personal effects of the technology in order to make it better serve our human purposes.'
> —Sherry Turkle, Abby Rockefeller Mauzé Professor of science, technology and society at MIT[1]

This book underlines a purpose-based approach to digital transformation as the key to success. While digital technologies have tremendous potential, digital transformation requires carefully avoiding the many painful transitions. Purposeful digital transformation outlines a way to do so. That is, the pursuit of purpose helps organizations and individuals avoid the transformative paradox—a phenomenon that underlines a slow growth in well-being despite tremendous technological achievements (revisit Chapter 1). To say that purpose has been a force central to all human lives will not be an overstatement. It is not feasible for an individual to work without a purpose. However, clarity and consciousness about one's purpose vary across individuals. The way one construes their own purpose differentiates the successful from the not-so-successful. Millions play cricket, sing, teach, act, preach or practise politics. Why do some succeed more than others?[2] Why do some rise to levels beyond others? Beyond other factors, a clear and well-thought-out purpose differentiates those who succeed. And identifying purpose requires one to think fundamentally, questioning self and existence.

For digital transformation, the assumption about 'who is an individual' lays the strongest foundation for success. Policymakers, business leaders, technology leaders and entrepreneurs are rethinking who is an individual, conceptualizing one scientifically as a neuroscientific being. This neuroscientific view of a wo(man) offers a unique lens. Notably, the view of an individual as a neuroscientific being is in sharp contrast with the widely-held assumptions about an individual as belonging to a certain species, cult, tribe, nation, religion, class, race and so on. All of these are aggregate and dated conceptualizations of an individual. A digital transformation built on these views of an individual is most likely to fail in comparison to one built with a neuroscientific focus. Using a neuroscientific approach, one must consider that a distinct set of neural networks and neuroscientific activations trigger emotions and behaviours. Activations in various brain regions characterize different individual processes—emotional, social, cognitive or decision-making.[3] Even very complex behaviours of an individual (e.g., risk, uncertainty, loss, happiness, trust, distrust, fear, biases, disgust and habits) map to brain regions.

When digitally transforming, this assumption about an individual opens up avenues for enhancing computational freedoms. The digital transformation manifests at the neuroscientific level. The computational freedom drives it, as seen in modern socializing, say using the power of mobile cameras and social networks like Instagram, WhatsApp and so on. Freedoms manifest in different ways. For example, digital social networks free people from the computations involved in driving vehicles, travelling, dressing up and so on, which are required for face-to-face meetings—the alternative to socializing digitally. At the organizational level, computational freedom (and costs due to lack of it) stack up. Profitable business models transforming industries emerge as firms capture these cost savings. For example, consider the transformation of mobility through ride-sharing.

Ride-sharing enhances asset efficiency, as many people share the same car (or another vehicle). Public sharing of buses, railways or airplanes has been common. While trust in public settings is easy, individuals lack trust when sharing private assets. The computational burden in private (or semi-public) settings is greater. People trust even strangers and share assets across the world, as enabled by Ola, Uber and Didi Global. Digital technologies computationally freed the parties (riders and drivers) from trust-related dilemmas. If two people are going in the same direction, the reason not to travel together is trust about others' intentions. Computations help predict the intentions of the other person. Specifically, digital transformation helps *predict* intentions using data (say, from social networks and a public trail of one's activity).[4] The value of stacking individual computational freedoms redefined mobility in the world.

The power of freedom (even in the non-digital age) may not be overemphasized. Freedom enables humans to evolve. For ages and across countries, the pursuit of freedom has stimulated the human mind. Across nations, freedom has been a cherished emotion, when subdued by foreign powers, ignorance, dogma and such. Indeed, the sense of freedom is real. The global revulsion for slavery underlines how strongly we value freedom. Slavery manifests even today and is a phenomenon whereby human beings are treated as objects, inventories to be carried and goods to be sold.[5] In older days, even children were not spared and were sold and separated from their families. Slavery created universal disgust. Neuroscientifically, it was a computational burden that was way too heavy. Because we sought freedom from this burden, humans across the globe stood up against slavery. For example, renowned

Indian poet, writer, composer, philosopher, playwright and social reformer Rabindranath Tagore expresses his wish for freedom that will transform his brethren living as slaves commanded by a foreign ruler.[6] Similarly, the founders of the US espoused freedom from political autocracy and royalty. The enormous human burden still shakes humanity's conscience and drives many to seek freedom (even for others). A continuous endeavour to enhance freedoms has led society to evolve to its current state.

For digital transformation, the historic journey to a modern, free society underlines the role of freedom as an even deeper pursuit at the neuroscientific level. To help managers master this aspect, the book has unveiled for you a central concept—*computational freedom*. The concept drives digital transformation and will continue to do so for generations to follow. And, to leverage it effectively, an organization has to transform purposefully, focusing on three facets:

1. The Instrumental Purpose: Building digital capabilities that enhance organizational performance with a focus on catalysing individual emotional goals
2. The Operational Purpose: Enacting the transformation through the extended DaWoGoMo© model, enhancing organization and individual explorations and exploitations.
3. The Existential Purpose: Clearly outline and communicate the computational freedoms enhanced by the artificially intelligent ecosystems developed through the transformation.

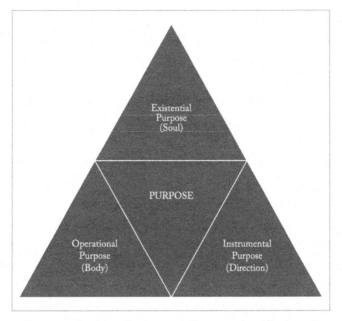

Figure 12: Purposes: Instrumental, Operational and Existential

Three Pillars of Digital Transformation: Direction, Body and Soul

Digital transformation isn't a singular pursuit; it thrives on a strong foundation built upon three interwoven purposes: instrumental, operational and existential. The three offer direction, body and soul to any organization's digital transformation, respectively (see Figure 12). First, instrumentally, an organization building digital capabilities advances its abilities, strategically moving in a specific direction. A focused *direction* for digital transformation helps realize synergistic areas for advancement. That is, the instrumental purpose helps choose the direction for the organization to follow, outperform competitors and overcome internal inertia. To successfully execute its instrumental purpose, any such transformation must build digital capabilities that help meet individual goals. Many digital capabilities have been built by enhancing an organization's online presence, transactional capabilities (say, by developing a UPI-based payment system) or enabling community information sharing (through online platforms). However, when it becomes an organization's singular pursuit, the instrumental purpose engenders a mechanical approach to digital transformation. If the organization focuses too heavily only on *what* to do, it misses out on the operational intricacies and the deep thinking that unleashes existential energy. This approach may be sufficient in times when there are limited options (or time) to think about *how* and *why*, and a comprehensive or creative transformation is not warranted.

In ordinary times, the second purpose—the operational purpose—guides an organization about *how* to transform. Mastering operational purpose entails understanding and changing the various organizational elements: Digital Architecture (Da), Work Systems (Wo), Governance (Go), Business Models (Mo) and Culture (collectively known as the extended DaWoGoMo© model). These changes form the *body* of a digital transformation. That is, even when the organization has decided the direction to evolve, it still needs to move and transform its digital architectures, work systems, decision-making authorities and business models, along with its culture. When this extended DaWoGoMo© model is operationalized with a focus on exploration and exploitation, the operational force is unleashed the most. Such an approach helps synchronize and integrate the five transformations as the organization applies the extended DaWoGoMo© to *learn,* evolving into a better version of itself. However, organizations solely focused on the operational purpose, neglecting the other two purposes, risk falling short. Such an organization lacks the clear direction required to identify high-performing capabilities. Also, it may lack the existential force—a clear and compelling purpose—that motivates stakeholders to actively champion the transformation.

Finally, a wise approach entails mastering the third purpose of digital transformation: the existential purpose. Each organization must ponder why it needs to digitally transform. The existential purpose offers a *soul* to the organization's digital transformation. It indicates that the digital transformation is driven by a reason, creating a strong zeal across stakeholders. Along with the direction and the body, the soul powers the digital transformation, not just internally, but with the outside world. It offers the organization an existential energy—one that has triggered the transformation of life and organization on the planet for ages. An organization succeeds the most when it champions the three purposes simultaneously, defining the transformation's direction (The Instrumental Purpose), its body (The Operational Purpose) and its soul (The Existential Purpose) (see Figure 13).

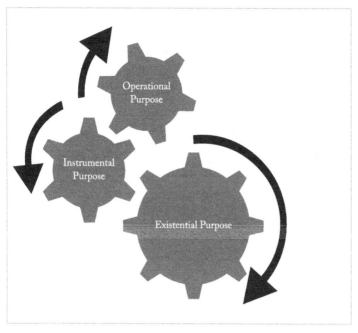

Figure 13: Three Intertwined Purposes

A Philosophical Message

No purpose becomes clear until it deeply questions the current 'truth'. The search for truth is an eternal human endeavour. Historically, people have been willing to even die in search of the truth. This is because the revelation of truth, at times, runs counter to the currently established practices or principles. Revelations disconfirming the geocentric model of the universe (thinking that the earth is the centre and everything revolves around us) led to large-scale turmoil in the western world.[7] Many technologies make us question our deeply held truths and rightly so. Advanced digital technologies, such as machine learning, deep learning and natural language processing, make us rethink our assumptions as they analyse and process large volumes of data much faster than humans. In many cases, their analysis is more effective. Using such technologies, it is now even possible to collect information from the brain and process it externally.[8] Consider the experiments on brain–computer interfaces at the Duke Medical Center. The centre reports the recording of neural activity from a macaque monkey's neocortex, triggering a robot arm.[9] When shown a banana, the monkey, with its biological arm tied and brain signals linked to a robot arm, moves its artificial limb to pick and eat the banana.[10] That is, the brain's intentions could be routed through digital technologies and linked with the robot arm. This could substitute for the need for a real arm. These technologies do, and should, make us question the truth about ourselves and life.

Who are we? The question gains prominence as the body's functionality has historically been linked with well-being, social respect and equality. Even today, in most organizations and jobs, the world order favours the person with a fully fit body. The digital technologies enabling the brain–computer interface open tremendous medical possibilities, so human beings with limited freedoms (say due to paralysed bodies) may enact their thoughts. A new world that respects everyone *equally* may emerge if we think differently about disabilities in our societies.

However, doing so requires challenging the current truths about them. In this context, a purposeful digital transformation requires enhancing freedoms for those who have limited physical functionalities. That is, while digitally transforming, you should think differently about human life and who we are. I have shared some answers in this book. However, the journey to find the truth is your personal endeavour.

This journey may lead you to new truths, freeing you from rigid assumptions. In the discussion on the chase in the jungle (see Chapter 11), the lion and deer are intertwined in the natural lifecycle, which itself is a rigid assumption. The assumption of a natural lifecycle underlines that either the lion will go hungry (and eventually die) or the deer will die. However, this zero-sum game prevailing in the jungle has been challenged in the human world. For example, having pet animals protects even the most vulnerable and weakest animals. Advanced technologies are now helping overcome the zero-sum game in the natural ecosystem. Notably, the digital transformation of our research has led to the growth of various alternate food materials, such as in-vitro or fake meat materials that taste close to meat and are now being served in restaurants.[11] Leaders such as Microsoft's Bill Gates find opportunities in these endeavours.[12] Whether such artificially intelligent ecosystems may be advanced to alleviate the zero-sum game in the natural kingdom and to what extent is a question that will be answered in the future. However, a huge focus on the environment, animals and climate, along with a billion-dollar opportunity, ensures that one has to watch out for such purposeful digital transformations that may shatter and transform rigid and reified assumptions.

Similarly, many digital transformations now challenge assumptions that outline the zero-sum trade-offs. These trade-offs imply that one may gain only when the other loses. Both nationally and internationally, many such phenomena are regularly observed with language being a common example. The world over, the languages differ. However, language is a two-person phenomenon. That is, if I speak a language the other person does not understand, there is a breakdown of communication. So, one has to learn another language to be able to communicate. Who would this person be—me or you? That is the genesis of the zero-sum game. One of the languages must be dominant. In China, it is Mandarin. So, Chinese moving abroad may have to learn English or the people coming to China must learn Mandarin to trade and transact. In India, the situation is more complex. India has a complex language landscape. There are over twenty official languages. I recall the first time I started working in a non-native state, touring the remote areas of southern India. I soon realized I needed a translator to communicate with people in many parts of *my* country. Now contrast this with my travelling within the US for many years. I could speak with everyone as I understood the language (English) and everyone else understood me. However, there was a tremendous cost upfront, largely for me. I had to invest time and effort in learning English, which was not my native language. If I were to spend that amount of effort on learning all the other languages, such as Marathi, Tamil, Telugu, Malayalam and so on, it would be very costly. How hard is it to solve the problem, and what is the problem?

I frame it as a *mass customization problem*. I recount my experience at the School of Business and Economics at the Universidad de Chile, Santiago. I was invited to deliver a keynote speech there. Before going, I did not have the time to research the school. Upon arrival at the university, I realized that the school was predominantly Spanish-speaking. I asked my host, almost feeling ashamed and guilty. He assured me, explaining that there were language-translator headphones that do the translation in real time. As I spoke to an auditorium full of people the next day, the interactions went great, without anyone bearing the cost of learning others' language. Technology had mass customized my words and sentences in Spanish and vice versa. Generically, one may see the value created by such automated translations in managing a congregation of large and diverse people. For example, during the 2018 Winter Olympics in South Korea, humanoid robots used AI-powered translation software to provide vital information to visitors and athletes. Mass customization works like a charm, resolving the potential of a zero-sum game. The question of who will learn the other language becomes irrelevant. A purposeful digital transformation challenges the 'truth', the zero-sum game.

Conclusion

Life on the planet evolves when leaders and followers pursue a purpose. Organizations like Google have differentiated themselves through their purpose, as highlighted by their mission statement: 'to organize the world's information and make it universally accessible and useful.'[13] New digital organizations are leaving a legacy much beyond what others even imagined. These demonstrate that the pursuit of purpose can transform lives on a large scale. As an example, India's Aadhaar transformed the lives of many, bringing many private and government services to their doorsteps.[14] The impact of purpose is far-reaching. Purposeful digital transformation often transforms societies and generations. For example, Aadhaar forms the backbone for large inclusion projects. The Prime Minister's Jan-Dhan Yojana (PMJDY) is one such financial inclusion programme launched in 2014.[15] On the first day of its launch, the programme led to the creation of 10 million bank accounts leveraging the Aadhaar IDs at very low costs and boasting 300 million new accounts in the next three years.[16] Beyond large organizations, digitally transforming *purposefully* is an urgency for most organizations today. Not following a purposeful approach is indicative of a very dark future lying ahead. The lives of many may get worse. That is, the digital transformation of the world, the digital age, may be very painful for many if the purpose is not manifested and well understood (revisit the discussion in Chapter 1). So, it is urgently required that we avoid being on the wrong side of digital transformation.

The Purpose

In this book, you have learnt that purpose represents an intelligent human organizational energy that manifests at three levels, as I outline the what, how and why of digital transformation. Digital technologies are an external force on human lives, in most cases. Yes, the creators of these technologies are humans. However, for most users, these are tools with features and functionalities. The creators of technologies have not always been the bearers of wisdom required to leverage technology for the greater good. However, they do offer a wonderful version of truth—a new possibility. Whether this is leveraged for good or leads to a dark future is unknown. While studying physics at college, I found nuclear energy to have the same dynamic. It is a source of abundant energy with the potential to transform lives positively; however, it did lead to large-scale destruction in the past. Similarly, I underline that intelligent organization and life may ensue only when a purpose-based approach is used to leverage digital technologies. If used appropriately, the model of purpose I outline will help overcome the transformative paradox, unleashing a creative and prosperous digital age.

Notes

Chapter 1: Digital Organization—The Digital Transformation of an Organization

1. David Shariatmadari, 'Physicist Michio Kaku: "We Could Unravel the Secrets of the Universe"', *Guardian*, 22 April 2023, https://www.theguardian.com/books/2023/apr/22/physicist-michio-kaku-we-could-unravel-the-secrets-of-the-universe.
2. Ganganatha Jha, 'Manusmriti with the Commentary of Medhatithi', verse 5.108, https://www.wisdomlib.org/hinduism/book/manusmriti-with-the-commentary-of-medhatithi/d/doc200491.html.
3. M. Burgess, 'Google's AI Has Written Some Amazingly Mournful Poetry', WIRED, 16 May 2016, http://www.wired.co.uk/article/google-artificial-intelligence-poetry; K. Finley, 'In the Future, Robots Will Write News That's All About You', WIRED, 6 March 2015, https://www.wired.com/2015/03/future-news-robots-writing-audiences-one/;

D. Rockmore, 'Human or Machine: Can You Tell Who Wrote These Poems?', NPR, 27 June 2016, https://www.npr.org/sections/alltechconsidered/2016/06/27/480639265/human-or-machine-can-you-tell-who-wrote-these-poems.

4 Companies such as Telenoid are creating telepresence robots that receive signals over the Internet and enact expressions using the signals. Also see B. McGee, 'Introducing Robotic Telepresence—The Creepy Side of Collaboration Technology', Vyopta, 19 January 2016, https://www.vyopta.com/blog/uc-industry/news/telepresence-robot/.

5 L. Buchen, 'Robot Makes Scientific Discovery All by Itself', WIRED, 2 April 2009, https://www.wired.com/2009/04/robotscientist/; D. Coldewey, 'AI Learns and Recreates Nobel-Winning Physics Experiment', TechCrunch, 16 May 2016, https://techcrunch.com/2016/05/16/ai-learns-and-recreates-nobel-winning-physics-experiment/.

6 F. Levy and R.J. Murnane, *The New Division of Labor: How Computers Are Creating the Next Job Market* (Princeton, NJ: Princeton University Press, 2012).

7 G. Colvin, *Humans Are Underrated: What High Achievers Know That Brilliant Machines Never Will* (New York, NY: Penguin Publishing Group, 2015); S. Pinker, *The Stuff of Thought: Language as a Window into Human Nature* (New York, NY: Penguin, 2007).

8 Wendy Sheehan Donnell, 'Bill Gates on the Next 40 Years in Technology', PCMag, 6 August 2022, https://www.pcmag.com/news/bill-gates-on-the-next-40-years-in-technology.

9 E. Brynjolfsson and B. Kahin, *Understanding the Digital Economy: Data, Tools, and Research* (Cambridge, MA: MIT Press, 2002).

10 'Facts and Figures 2023', ITU, accessed 17 June 2024, https://www.itu.int/itu-d/reports/statistics/2023/10/10/ff23-internet-use/.

11 Saif M. Khan and Alexander Mann, 'AI Chips: What They Are and Why They Matter', Center for Security and Emerging Technology, April 2020, https://cset.georgetown.edu/publication/ai-chips-what-they-are-and-why-they-matter/.

12 J.W. Schwank and M. DiBattista, 'Oxygen Sensors: Materials And Applications', MRS Bulletin, 24(06) (1999): 44–48.

13 E. Vitz and H. Chan, 'LIMSport VII. Semiconductor Gas Sensors as GC Detectors and "Breathalyzers"', *J. Chem. Educ.* 72(10) (1995): 920.

14 M. Matson and J. Dozier, 'Identification of Subresolution High Temperature Sources Using a Thermal IR Sensor', *Photogrammetric Engineering and Remote Sensing*, 47(9) (1981): 1311–18.

15 S. Costello, 'The 6 Sensors in an iPhone', Lifewire, 4 February 2020, https://www.lifewire.com/sensors-that-make-iphone-so-cool-2000370.

16 P. Domingos, *The Master Algorithm: How the Quest for the Ultimate Learning Machine Will Remake Our World* (New York, NY: Basic Books, 2015).

17 Jeffrey Dunn, 'Introducing FBLearner Flow: Facebook's AI Backbone', Engineering at Meta, 9 May 2016, https://engineering.fb.com/2016/05/09/core-data/introducing-fblearner-flow-facebook-s-ai-backbone/.

18 M. Ford, *Rise of the Robots: Technology and the Threat of a Jobless Future* (New York, NY: Basic Books, 2015).

19 C. Shannon, 'A Mathematical Theory of Communication', *Bell System Technical Journal*, 27 July and October (1948): 379–423, 623–656; C.E. Shannon, 'A Mathematical Theory of Communication', *ACM SIGMOBILE Mobile Computing and Communications Review*, 5(1) (2001): 3–55.

20 Interested readers are advised to refer to works on data mining, deep learning and data science (for example, Provost and Fawcett 2013).

21 S. Wolfram, 'Latest Perspectives on the Computation Age', Stephen Wolfram Blog, 11 October 2012, http://blog.stephenwolfram.com/2012/10/latest-perspectives-on-the-computation-age/.

22 C. Metz, 'Google's AlphaGo Continues Dominance With Second Win in China', WIRED, 25 May 2017, from https://www.wired.com/2017/05/googles-alphago-continues-dominance-second-win-china/.

23 I. Fister, K. Ljubič, P.N. Suganthan and M. Perc, 'Computational Intelligence in Sports: Challenges and Opportunities within a New Research Domain', *Applied Mathematics and Computation*, 262 (2015): 178–186; R. Kurzweil, *The Age of Spiritual Machines: When Computers Exceed Human Intelligence* (New York, NY: Penguin, 2000).

24 E. Brynjolfsson and A. McAfee, *The Second Machine Age: Work, Progress, and Prosperity in a Time of Brilliant Technologies* (New York, NY: WW Norton & Company, 2014); S. Kroft, 'Are Robots Hurting Job Growth?', CBS News, 13 November 2013, from https://www.cbsnews.com/news/are-robots-hurting-job-growth-13-01-2013/.

25 L. Elliott, 'Rising Inequality Threatens World Economy, Says WEF', *Guardian*. 11 January 2017, http://www.theguardian.com/business/2017/jan/11/inequality-world-economy-wef-brexit-donald-trump-world-economic-forum-risk-report.

26 'The Freedom Dividend', Yang2020, https://2020.yang2020.com/policies/the-freedom-dividend/. Also see Chapter 7.

27 TED, 'Erik Brynjolfsson: The Key to Growth? The Race with Machines', YouTube video, 23 April 2013, https://youtu.be/sod-eJBf9Y0?si=1NrIQRWNiZbQCwaZ.

28 Catherine Clifford, 'Bill Gates: AI Is Like Nuclear Energy— Both Promising and Dangerous', CNBC, 26 March 2019,

https://www.cnbc.com/2019/03/26/bill-gates-artificial-intelligence-both-promising-and-dangerous.html.
29 See Colvin (2015).
30 See Colvin (2015) for the analogy.
31 Dr Hawking was a physicist who worked in the fields of general relativity, black holes and the origins of the universe. His contributions to physics earned him the Presidential Medal of Freedom, which is the highest civilian honour in the US.
32 Lauren Walker, 'Stephen Hawking Warns Artificial Intelligence Could End Humanity', *Newsweek*, 14 May 2015, https://www.newsweek.com/stephen-hawking-warns-artificial-intelligence-could-end-humanity-332082.
33 The timelines may be debated across different civilizations. For example, the Indus Valley civilization—the second oldest human civilization (after Mesopotamia)—outlines the presence of wheel-based carts around 7500 BCE. However, the core argument is not specific to the exact timelines of these inventions.
34 E. Brynjolfsson and A. McAfee, *The Second Machine Age: Work, Progress, and Prosperity in a Time of Brilliant Technologies*. Also see Waymo, https://waymo.com/
35 RPA involves the use of robots to automate work.
36 This represents a model for creating healthcare services that involve patients directly as providers of crucial information, such as scheduling appointments, uploading vitals and tests, etc. For more, see my research: R. Aljafari, F. Soh, P. Setia and R. Agarwal, 'The Local Environment Matters: Evidence from Digital Healthcare Services for Patient Engagement', *Journal of the Academy of Marketing Science* (2023): 1–23, https://doi.org/10.1007/s11747-023-00972-0.
37 Jonathan Becher, 'Separating the Digital Revolutionaries from the Reactionaries', *Forbes*, 21 December, 2016, https://

www.forbes.com/sites/sap/2016/12/21/separating-the-digital-revolutionaries-from-the-reactionaries/?sh=616dce7d1733.

Chapter 2: Digital Transformation of Organization: The Scientific Foundations

1 Carl Sagan, *Broca's Brain: Reflections on the Romance of Science* (New York: Random House Publishing Group, Kindle Edition, 2011). (Kindle Locations 344–346). Also see: Jack Hassard, 'The Art of Teaching Science: Science is a Way of Thinking: So, Why Do We Try and Standardize it?', NEPC, 11 March 2014, https://nepc.colorado.edu/blog/science-way-thinking.
2 Such as Arduino and Raspberry Pi boards and 3D printers.
3 N.G. Carr, 'IT Doesn't Matter', *Harvard Business Review* 81(5) (2003): 41–49.
4 Ibid.
5 Muhammad Ali Chaudhry and Emre Kazim, 'Artificial Intelligence in Education (AIEd): A High-Level Academic and Industry Note 2021', NIH, 7 July 2021, https://www.ncbi.nlm.nih.gov/pmc/articles/PMC8261391/; J.R. Varma, S. Fernando, B.Y. Ting, S. Aamir and R. Sivaprakasam, 'The Global Use of Artificial Intelligence in the Undergraduate Medical Curriculum: A Systematic Review', NIH, 30 May 2023, https://www.ncbi.nlm.nih.gov/pmc/articles/PMC10309075/
6 D. Leonard-Barton and J.J. Sviokla, 'Putting Expert Systems to Work', *Harvard Business Review*, March 1988, https://hbr.org/1988/03/putting-expert-systems-to-work.
7 J.E. Triplett, 'The Solow Productivity Paradox: What Do Computers Do to Productivity?', *The Canadian Journal of Economics/Revue Canadienne d'Economique*, 32(2) (1999): 309–334.
8 In-depth follow-up analysis revealed some errors in estimation, leading to the resolution of the paradox and the conclusion

that information technologies have positive impacts on an organization's performance. See E. Brynjolfsson and L. Hitt, 'Paradox Lost? Firm-Level Evidence on the Returns to Information Systems Spending', *Management Science*, 42(4) (1996): 541–558.

9   It is important to underline that innovation is different from invention, and the former defines organizational purpose. Simplistically, discovering a novel technology is an invention, and applying it in a new way–to organize–is innovation. In one of the conversations, the MD of a leading Silicon Valley firm once told me that the organization has rarely invented and most of its products are essentially organizational innovations; B. Walker, 'Innovation vs. Invention: Make the Leap and Reap the Rewards', WIRED, January 2015, https://www.wired.com/insights/2015/01/innovation-vs-invention/.

10 It is important to note that digital transformations are gaining significance because of the novel advances in the technology space—neural network activation functions, numerous patented technologies for personalized recommendations, peer-to-peer or merchant payments, fund transfers, chatbots, predictive analytics, autonomous vehicles and industrial automation, among others. See Chapter 1 as well.

11 This does not imply that there were no new technologies developed. Instead, this emphasizes the relative role of innovation vis-à-vis new technological inventions by these organizations.

12 Primarily, because science does not mean natural sciences (physics, chemistry, etc.).

13 Brian Greene, 'Put a Little Science in Your Life', *New York Times*, 1 June 2008, https://www.nytimes.com/2008/06/01/opinion/01greene.html.

14 Karl E. Weick, 'Theory Construction as Disciplined Imagination,' *The Academy of Management Review* 14, no. 4 (1989): 516–31. https://doi.org/10.2307/258556.
15 T. D'Augustino, 'Helping Youth Succeed in Science—Part 6: Using Mathematics and Computational Thinking', Michigan State University, 17 November 2015, https://www.canr.msu.edu/news/helping_youth_succeed_in_science_part_6_using_mathematics_and_computational.
16 TED, 'Murray Gell-Mann: Beauty and Truth in Physics', YouTube video, 7 December 2007, https://www.youtube.com/watch?v=UuRxRGR3VpM&t=129s.

Chapter 3: The Instrumental Purpose of the Organization's Digital Transformation

1 Sundar Pichai, 'Investing in India's Digital Future', The Keyword, 13 June 2020, https://blog.google/inside-google/company-announcements/investing-in-indias-digital-future/.
2 It is important to note that performance is not necessarily financial (measured in rupees, renminbi or US dollars). An organization (say, a hospital) may perform well when it saves lives. Similarly, organizations may pursue goals related to creating intelligent minds (say, a university) or enhancing the religious spirit (temple, church or mosque).
3 Nicole Goodkind, 'The Stock Market Is Dominated by Just a Handful of Companies. The Biden Administration Is Worried', CNN, 25 July 2023, https://edition.cnn.com/2023/07/25/investing/premarket-stocks-trading/index.html.
4 D.J. Teece, G. Pisano and A. Shuen, 'Dynamic Capabilities and Strategic Management', *Strategic Management Journal*, 18(7) (1997): 509–33.

5 M. Porter, Competitive Strategy (New York, Free Press, 1980).
6 E. T. Penrose, (2009), *The Theory of the Growth of the Firm* (Oxford University Press), J. Barney, (1991), 'Firm Resources and Sustained Competitive Advantage', *Journal of Management*, *17*(1), 99–120; M.A. Peteraf, (1993), 'The Cornerstones of Competitive Advantage: A Resource-Based View', *Strategic Management Journal*, *14*(3), 179–91.
7 See Teece et al. (1997) above.
8 C. Scarre, *Smithsonian Timelines of the Ancient World* (Washington, D.C.: Smithsonian Institution, 1993).
9 Ibid.
10 Ibid.
11 Bill Gates, 'The Age of AI Has Begun', GatesNotes, 21 March 2023, https://www.gatesnotes.com/The-Age-of-AI-Has-Begun.
12 R.R. Nelson and S.G. Winter, 'The Schumpeterian Tradeoff Revisited', *American Economic Review*, 72(1) (1982): 114–32.
13 Teece et al. (1997): 509–33.
14 P.A. Pavlou and O.A. El Sawy, 'From IT Leveraging Competence to Competitive Advantage in Turbulent Environments: The Case of New Product Development', *Information Systems Research*, 17(3) (2006): 198–227.
15 P. Setia, and P.C. Patel, 'How Information Systems Help Create OM Capabilities: Consequents and Antecedents of Operational Absorptive Capacity', *Journal of Operations Management*, 31(6) (2013): 409–31.
16 P. Setia, V. Venkatesh and S. Joglekar, 'Leveraging Digital Technologies: How Information Quality Leads to Localized Capabilities and Customer Service Performance', *MIS Quarterly* (2013): 565–90.
17 Ibid.
18 See Setia and Patel (2013) above.

19 See Gates (2023), https://www.gatesnotes.com/The-Age-of-AI-Has-Begun.
20 Jenna Goudreau, 'IBM CEO Predicts Three Ways Technology Will Transform the Future of Business', *Forbes*, 8 March 2013, https://www.forbes.com/sites/jennagoudreau/2013/03/08/ibm-ceo-predicts-three-ways-technology-will-transform-the-future-of-business/?sh=915e6df6a732.
21 A major airline found that there was a ten-minute gap between estimated and actual times for 10 per cent of the flights, and a five-minute deviation for 30 per cent of the flights.
22 A. McAfee and E. Brynjolfsson, 'Big Data: The Management Revolution', *Harvard Business Review*, October 2012, https://hbr.org/2012/10/big-data-the-management-revolution.
23 See Gates (2023), https://www.gatesnotes.com/The-Age-of-AI-Has-Begun.
24 See Setia and Patel (2013) and Setia et al. (2013) above.
25 See D.J. Teece, G. Pisano and A. Shuen (1997) above.

Chapter 4: Visualizing and Conceptualizing Digital Capabilities

1 'Technology Should Drive More Inclusive Economic Growth: Satya Nadella', Microsoft News, https://news.microsoft.com/en-in/features/technology-should-drive-inclusive-economic-growth-satya-nadella/.
2 'Companies Have to Build Their Own Tech Capability: Microsoft CEO Satya Nadella', e4m, 25 February, 2020, https://www.exchange4media.com/digital-news/companies-have-to-build-their-own-tech-capability-microsoft-ceo-satya-nadella-102929.html.
3 See Pichai (2020), https://blog.google/inside-google/company-announcements/investing-in-indias-digital-future/.

4   N.G. Carr, 'IT Doesn't Matter', *Harvard Business Review* 81(5) (2003): 41–49.
5   This is not intended to undermine the value of other perspectives. Instead, it is an assertion that the evolutionary way works, and I have researched its applicability for building advanced digital capabilities in large and small organizations.
6   I will delve much deeper into the individual evolutionary dynamics later, in Chapter 11.
7   T. Gerken, 'Bill Gates: AI Is Most Important Tech Advance in Decades', BBC.com, 22 March 2023, https://www.bbc.com/news/technology-65032848.
8   Pankaj Setia, Viswanath Venkatesh, and Supreet Joglekar (2013), 'Leveraging IS for Creating a Customer-Centric Organization: How Information Quality Leads to Localized Capabilities and Service Performance', *MIS Quarterly*, Vol. 37, Iss. 2, 565–90.
9   This would have been an argument for the moderating impact.

Chapter 5: Operationalizing Digital Transformation

1   Dr C. Block, '12 Reasons Your Digital Transformation Will Fail', *Forbes*, 16 March 2022, https://www.forbes.com/sites/forbescoachescouncil/2022/03/16/12-reasons-your-digital-transformation-will-fail/?sh=29ebc7a11f1e.
2   Parc of vehicles defines the number of vehicles on road (accounting for the ones that go off the road each year). Also see M. Kahl, 'Ford on Why Smart Mobility Is Essential for Smart Cities', *Automotive World*, 12 June 2018, https://www.automotiveworld.com/articles/ford-smart-mobility-essential-smart-cities/.
3   'Ford Smart Mobility Llc Established to Develop, Invest in Mobility Services; Jim Hackett Named Subsidiary Chairman',

Ford, 11 March 2016, https://media.ford.com/content/fordmedia/fna/us/en/news/2016/03/11/ford-smart-mobility-llc-established--jim-hackett-named-chairman.html.

4 'Ford Dynamic Shuttle Service Moves from Experiment to Pilot, Providing Point-To-Point Shuttle Rides for Employees', Ford, 10 December 2015, https://media.ford.com/content/fordmedia/fna/us/en/news/2015/12/10/ford-dynamic-shuttle-service-moves-from-experiment-to-pilot.html.

5 B. Morgan, 'Companies That Failed at Digital Transformation and What We Can Learn from Them', *Forbes*, 30 September 2019, https://www.forbes.com/sites/blakemorgan/2019/09/30/companies-that-failed-at-digital-transformation-and-what-we-can-learn-from-them/?sh=2a512ed2603c.

6 M.L. Markus, S. Axline, D. Petrie and C. Tanis, 'Learning from Hershey: Enterprise Resource Planning Implementation and Its Lessons', *Harvard Business Review*, 78(4) (2000): 46–55.

7 Source: United States Government Accountability Office. (2014). Healthcare.gov: Ineffective Planning and Oversight Practices Underscore the Need for Improved Contract Management.

8 Ibid.

9 S. Kien, C. Soh, P. Weil, and Y. Chong, (2015), 'Rewiring the Enterprise for Digital Innovation: The Case of DBS Bank', *Nanyang Business School, the Asian Business Case Center*.

10 'Introducing DBS PayLah!', DBS.com, https://www.dbs.com.sg/personal/deposits/pay-with-ease/dbs-paylah.

11 'DBS IDEAL—Online Internet Banking, DBS.com, https://www.dbs.com/in/sme/day-to-day/ways-to-bank/online-banking-ideal.

12 R. Balakrishnan, "DBS's Chilli Paneer Campaign: Was the Result Worth the Hard Work?', 4 February 2015, https://

economictimes.indiatimes.com/dbss-chilli-paneer-campaign-was-the-result-worth-the-hard-work/articleshow/46107746.cms?from=mdr.

Chapter 6: Transforming Digital Architecture

1 'Global Leaders Usher in a New Era of Digital Cooperation for a More Sustainable, Equitable World', UNDP.org, 22 September, 2022, https://www.undp.org/press-releases/global-leaders-usher-new-era-digital-cooperation-more-sustainable-equitable-world.
2 Ibid.
3 M.R. Vitale, 'American Hospital Supply Corp.: The ASAP System (A)', Harvard Business Publishing Education, 10 July 1985, https://hbsp.harvard.edu/product/186005-PDF-ENG?activeTab=include-materials.
4 P. Weill and S. Aral, 'Generating Premium Returns on Your It Investments', *MIT Sloan Management Review*, (2006).
5 'Digital Innovation and Portfolios', MIT Center for Information Systems Research, https://cisr.mit.edu/content/classic-topics-digital-innovation-and-portfolios. Also, see Weill and Aral (2006) above.
6 Franck Soh and Pankaj Setia, 'The Impact of Dominant IT Infrastructure in Multi-Establishment Firms: The Moderating Role of Environmental Dynamism', *Journal of the Association for Information Systems*, 23(6) (2022): 1603–33. DOI: 10.17705/1jais.00773, available at: https://aisel.aisnet.org/jais/vol23/iss6/2.
7 J.R. Galbraith, 'Organization Design: An Information Processing View', *Interfaces*, 4(3) (1974): 28–36.
8 The links between individuals and organizations' information processing are deeply embedded and form the foundations of

digital transformation. Also see Part 5 (e.g., Chapter 14) where I discuss why digital transformation is so existentially vital.

9 As I will explain later, digital architecture transformation is linked with the other transformations (of work, governance and business models). Therefore, digital architecture transformation has to be in line with the overall digital transformation of the organization.

10 Le Corbusier, *Towards an Architecture*, 4th ed. (Routledge, 2017), 3, eBook ISBN 9781315303673.

11 Vitruvius, *Ten Books on Architecture*. Translated by Ingrid D. Rowland (Cambridge: Cambridge University Press, 1999).

12 R.L. Nolan, K. Porter and C. Akers, 'Cisco Systems Architecture: ERP and Web-enabled IT', Harvard Business School Case 301–099, March 2001 (Revised November 2005).

13 Google, 'Architecting Disaster Recovery for Cloud Infrastructure Outages', *Google Cloud Architecture Center*, https://cloud.google.com/architecture/disaster-recovery.

14 S.K. Sia, C. Soh, P. Weill and Y. Chong, 'Rewiring the Enterprise for Digital Innovation: The Case of DBS Bank', Harvard Business Publishing Education, 2 June 2015, https://hbsp.harvard.edu/product/NTU071-PDF-ENG?activeTab=overview.

15 T. Davenport, (2014), *Big Data at Work: Dispelling the Myths, Uncovering the Opportunities* (Boston, MA: Harvard Business School Publishing Corporation).

16 F. Provost, and T. Fawcett, (2013), *Data Science for Business: What You Need to Know about Data Mining and Data-Analytic Thinking* (O'Reilly Media, Inc.).

17 'e-NAM Overview', e-NAM, https://www.enam.gov.in/web/.

18 P&G Connect+Develop, https://www.pgconnectdevelop.com/.

19 Counts as of October 2016 from SourceForge.net, the largest community of OSS projects on the Internet.

20 'Which Are the Most Popular Web Servers?', Stackscale, https://www.stackscale.com/blog/top-web-servers/, accessed November 2023.
21 E.S. Raymond, *The Cathedral and the Bazaar* (Sebastopol, CA: O'Reilly Media, 1999), accessed March 2024, https://lists.gnu.org/archive/html/groff/2021-11/pdfRt8tRop5yy.pdf.

Chapter 7: Transforming Work

1 S. Majumdar, K. Gopalan, 'We Will Invest More in India and from India to the World,' says Google boss Sundar Pichai, 19 February 2023, https://www.businesstoday.in/magazine/interview/story/bt-exclusive-we-will-invest-more-in-india-and-from-india-to-the-world-says-google-boss-sundar-pichai-369017-2023-02-05.
2 The concept is no longer limited to the realm of sci-fi movies, it is real for now. See, 'Live From Surgery', Liberty Science Center, https://lsc.org/education/educators/live-from-surgery
3 R.G. Zand, 'In Global First, Shaare Zedek Spine Surgeon Combines Augmented Reality with Robotics', *Times of Israel*, 30 August 2023, https://www.timesofisrael.com/in-global-first-shaare-zedek-spine-surgeon-combines-augmented-reality-with-robotics/.
4 P. Milgrom and J. Roberts, 'Complementarities and Fit Strategy, Structure, and Organizational Change in Manufacturing', *Journal of Accounting and Economics*, 19(2–3) (1995): 179–208.
5 Pollution in large cities is a concern, and in many Asian cities, it is life-threatening. See P. Agarwal, 'You Are Losing 11.9 Years of Your Life to Toxic Air in Delhi', 30 August 2023, https://timesofindia.indiatimes.com/city/delhi/you-are-losing-11-9-years-of-your-life-to-toxic-air-in-delhi/articleshow/103188256.cms.

6 See Milgrom and Roberts (1995) above.
7 D.D. Cremer and G. Kasparov, 'AI Should Augment Human Intelligence, Not Replace It', *Harvard Business Review*, 18 March 2021, retrieved from https://www.kasparov.com/ai-should-augment-human-intelligence-not-replace-it-harvard-business-review-march-18-2021/.
8 See Gates (2023), https://www.gatesnotes.com/The-Age-of-AI-Has-Begun.
9 'ICICI Lombard and Skit.ai Partner to Launch a First-of-Its-Kind AI-Powered Digital Voice Agent', E-CIO, 14 September 2022, https://cio.eletsonline.com/news/icici-lombard-and-skit-ai-partner-to-launch-a-first-of-its-kind-ai-powered-digital-voice-agent/69777/.
10 Source(s): https://adage.com/datacenter/datapopup.php?article_id=317270.
11 J. Vanian, 'Bill Gates Says A.I. Could Kill Google Search and Amazon as We Know Them', CNBC, 22 May 2023, https://www.cnbc.com/2023/05/22/bill-gates-predicts-the-big-winner-in-ai-smart-assistants.html.
12 Thomas H. Davenport, (2014), *Big Data at Work: Dispelling the Myths, Uncovering the Opportunities* (Boston, MA: Harvard Business School Publishing Corporation); Thomas H. Davenport and Jinho Kim, *Keeping Up with the Quants: Your Guide to Understanding and Using Analytics* (Harvard Business Review Press, 2013).
13 ABB, 'ABB's Robot YuMi Takes Center Stage in Pisa, Conducts Andrea Bocelli and Lucca Symphony Orchestra', YouTube video, 13 September 2017, https://www.youtube.com/watch?v=fohc1Qg-rQU.
14 'After Machines Fail, "Rat Miners" to Help Rescue 41 Men Stuck in Uttarkashi Tunnel', *Economic Times*, 27 November 2023, https://economictimes.indiatimes.com/news/india/

after-machines-fail-rat-miners-to-help-rescue-41-men-stuck-in-uttarkashi-tunnel/articleshow/105534580.cms.

15 So, a person standing still and carrying a lot of weight may not be considered working.

16 L. Selden and I.C. MacMillan, (2006), 'Manage Customer-Centric Innovation-Systematically', *Harvard Business Review*, *84*(4), 108.

17 M. Solomon, 'How TUMI Transformed Its Customer Service to Be as Bulletproof as Its Luggage', 29 December 2019, https://www.forbes.com/sites/micahsolomon/2018/12/29/how-tumi-transformed-its-customer-service-to-be-as-bulletproof-as-its-luggage/?sh=615194a650e5.

18 See Gates (2023), https://www.gatesnotes.com/The-Age-of-AI-Has-Begun.

19 R.T. Rust, C. Moorman and G. Bhalla, 'Rethinking Marketing', *Harvard Business Review*, 88(1/2) (2010): 94–101.

20 D.L. Goodhue and R.L. Thompson, 'Task-Technology Fit and Individual Performance', *MIS Quarterly*, 19, No. 2 (1995): 213–36.

21 A.R. Dennis, R.M. Fuller and J.S. Valacich, 'Media, Tasks, and Communication Processes: A Theory of Media Synchronicity', *MIS Quarterly*, 32, No. 3 (2008): 575–600.

22 R.L. Daft and R.H. Lengel, 'Information Richness: A New Approach to Managerial Information Processing and Organization Design', in *Research in Organizational Behavior*, eds. B. Staw and L. Cummings (Greenwich, Connecticut: JAI Press, 1984): 191–233.

23 ORTEC, 'Bulk Liquids Replenishment: Inventory Routing', https://ortec.com/en/solutions/bulk-liquids-replenishment/inventory-routing, Automated Stock Replenishment, https://ortec.com/en-us/topics/automated-stock-replenishment.

24 J.R. Hackman and G.R. Oldham, *Work Redesign* (Reading, MA: Addison–Wesley, 1980).

25 JCM underlines the role of five job characteristics: task significance, task identity, skill variety, autonomy and feedback as determinants of job satisfaction.
26 This is evident in the way we organize our lives as well. People have a daily morning routine. We brush our teeth every morning and so on. Professional organizations' routines may be more complex.
27 T.H. Davenport, J.G. Harris and R. Morison, (2010), *Analytics at Work: Smarter Decisions, Better Results*, Harvard Business Press.
28 H.J. Watson, B.H. Wixom, J.A. Hoffer, R. Anderson-Lehman and A.M. Reynolds, (2006), 'Real-Time Business Intelligence: Best Practices at Continental Airlines', *Information Systems Management*, 23(1), 7-18.
29 T.H. Davenport, J.G. Harris and R. Morison, (2010), *Analytics at Work: Smarter Decisions, Better Results*.
30 T. Allas, J. Maksimainen, J. Manyika and N. Singh, 'An Experiment to Inform Universal Basic Income', McKinsey, 15 September 2020, https://www.mckinsey.com/industries/social-sector/our-insights/an-experiment-to-inform-universal-basic-income.
31 M. Ferdosi, T. MacDowell, W. Lewchuk and S. Ross, 'On How Ontario Trialed Basic Income', UNESCO, 24 February 2022, https://en.unesco.org/inclusivepolicylab/analytics/how-ontario-trialed-basic-income.
32 'The Freedom Dividend', Yang2020, https://2020.yang2020.com/policies/the-freedom-dividend/.
33 'Switzerland's Voters Reject Basic Income Plan', BBC, 5 June 2016, https://www.bbc.com/news/world-europe-36454060#:~:text=Swiss%20voters%20have%20overwhelmingly%20rejected,whether%20they%20worked%20or%20not.

34 D. Varshney et al., 'India's COVID-19 Social Assistance Package and Its Impact on the Agriculture Sector', ScienceDirect, 9 January 2021, https://www.sciencedirect.com/science/article/pii/S0308521X21000020#fn0005.

Chapter 8: Transforming Governance

1 N. Tichy and W. Bennis, 'Making Judgment Calls', *Harvard Business Review*, October 2007, https://hbr.org/2007/10/making-judgment-calls.
2 S.K. Sia, C. Soh, P. Weill and Y. Chong, 'Rewiring the Enterprise for Digital Innovation: The Case of DBS Bank', 2 June 2015, https://hbsp.harvard.edu/product/NTU071-PDF-ENG
3 M.C. Jensen and W.H. Meckling, 'Theory of the Firm: Managerial Behavior, Agency Costs and Ownership Structure', *Journal of Financial Economics*, 3(4) (1976): 305–60.
4 Evidence of this has been observed at the brain activity levels. In our daily lives, this involves individuals evaluating their previous comprehension and general world knowledge. For example, Ferstl et al. (2005) examined the comprehension of language to assess utterances and sentences. In Ferstl's experiment, twenty participants listened to thirty-two short stories. As they were listening to the stories, fMRI was used to scan their entire head using a spin-echo sequence. The study found that information inconsistent with the global emotional state of the individual led to activations in the ventromedial prefrontal cortex and the extended amygdaloid complex. That is, individuals judged the emotional inconsistencies in the form of activations in certain brain regions. Further, individual attempts to integrate the inconsistent information led to activations in the dorsal frontomedial cortex (Brodmann's area

8/9). Furthermore, the brain activations were much stronger for inconsistencies that were emotional in nature, rather than inconsistent chronological or temporal information. See E.C. Ferstl, M. Rinck and D.Y.V. Cramon, 'Emotional and Temporal Aspects of Situation Model Processing during Text Comprehension: An Event-Related fMRI Study', *Journal of Cognitive Neuroscience*, 17(5) (2005): 724–39.

5 R.G. Sandell, (1968), 'Effects of Attitudinal and Situational Factors on Reported Choice Behavior', *Journal of Marketing Research*, 5(4), 405–08.

6 As I will argue later (in Chapter 12), judgements have an existential purpose, as emotions are core to human life.

7 This is a simplified view of governance. A complex aspect, governance may entail complex and nuanced variations in the way various organizations, projects or communities are managed. For example, formal and informal mechanisms may span behaviour control, clan control, outcome control and self-control. Many previous works, including mine, have studied the complexities of governance. Also see:

O. Treib, H. Bähr and G. Falkner, 'Modes of Governance: Towards a Conceptual Clarification', *Journal of European Public Policy*, 14(1) (2007): 1–20.

V. Sambamurthy and R.W. Zmud, 'Arrangements for Information Technology Governance: A Theory of Multiple Contingencies', *MIS Quarterly*, 23, No. 2 (1999): 261–90.

V. Choudhury and R. Sabherwal, 'Portfolios of Control in Outsourced Software Development Projects', *Information Systems Research*, 14(3) (2003): 291–314.

V. Venkatesh, T.A. Sykes, A. Rai and P. Setia, 'Governance and ICT4D Initiative Success: A Longitudinal Field Study of Ten Villages in Rural India', *MIS Quarterly*, 43(4) (2019): 1–24.

W.G. Ouchi, 'A Conceptual Framework for the Design of Organizational Control Mechanisms', *Management Science*, 25(9) (1979): 833–48.

8. See Sia et al. (2015) https://hbsp.harvard.edu/product/NTU071-PDF-ENG.
9. A simplistic description of ratifying and monitoring, and initiating and implementing, as control and monitoring activities, respectively. This conceptualization of decision rights was pioneered by Michael Jensen and William Meckling in 1976 and has proven its applicability across various domains. See Michael, C. Jensen, and William, H Meckling, 'Theory of the Firm: Managerial Behavior, Agency Costs and Ownership Structure', *Journal of Financial Economics*, October, 1976, V. 3, No. 4, 305–60.
10. F. Hansen, 'Psychological Theories of Consumer Choice', *Journal of Consumer Research*, 3(3) (1976): 117–42, https://doi.org/10.1086/208660.
11. J.G. March, 'Bounded Rationality, Ambiguity, and the Engineering of Choice', *Bell Journal of Economics*, 9(2) (1978): 587–608, https://doi.org/10.2307/3003600.
12. H.E. Krugman, 'Brain Wave Measures of Media Involvement', *Journal of Advertising Research*, 11(1) (1971): 3–9.
13. 'Motor Vehicle Crashes: A Leading Cause of Death in Younger People', Connecticut Department of Public Health, https://portal.ct.gov/-/media/Departments-and-Agencies/DPH/dph/hems/injury/NPHWFactSheetMOTORVEHICLEpdf.pdf.
14. See Hansen 1976 above.
15. https://waymo.com/.
16. 'Hollywood Actors and Writers' Strike Called Off: What We Know about the Deal So Far', *Indian Express*, 10 November 2023, https://indianexpress.com/article/explained/explained-global/hollywood-actors-writers-strike-deal-explained-9020526/.

17 'Shedding Light on AI Bias with Real World Examples', IBM, 16 October 2023, https://www.ibm.com/blog/shedding-light-on-ai-bias-with-real-world-examples/.
18 M. Zetlin, 'Bill Gates Says We're Witnessing a "Stunning" New Technology Age. 5 Ways You Must Prepare Now', INC, 24 March 2023, https://www.inc.com/minda-zetlin/bill-gates-says-were-witnessing-a-stunning-new-technology-age-5-ways-to-prepare.html.

Chapter 9: Transforming Business Model

1 'Alan Kay', TED, https://www.ted.com/speakers/alan_kay
2 Direct network effects also manifest on digital platforms, as the presence of more consumers benefits other consumers. For example, direct network effect manifests at Amazon, Flipkart and other similar online retail platforms (as more consumers add more reviews, benefiting other customers). Also, one may think about this direct network effect manifesting in social media platforms such as Facebook, where the presence of more friends motivates one to join.
3 For those not familiar, a tandoor is a clay oven used in India, Pakistan and some parts of the Middle East to make rotis, naans and other food items.
4 For all references to value throughout the text, one may assume that value is calculated by discounting future returns at an appropriate rate.
5 G.R. Carroll, 'A Stochastic Model of Organizational Mortality: Review and Reanalysis', *Social Science Research*, 12(4) (1983): 303–29.
6 Z. Cheng and B.R. Nault, 'Industry Level Supplier-Driven IT Spillovers', *Management Science*, 53(8) (2007): 1199–216.

7   M. Saunders, 'Strategic Purchasing and Supply Chain Management', Pearson, (1997); G.P. Cachon and M. Fisher, 'Supply Chain Inventory Management and the Value of Shared Information', *Management Science*, 46(8) (2000): 1032–48.

8   See Cachon and Fisher (2000) above; H.L Lee, V. Padmanabhan and S. Whang, 'Information Distortion in a Supply Chain: The Bullwhip Effect', *Management Science*, 43(4) (1997): 546–58.

9   K.E. Bourland, S.G. Powell and D.F. Pyke, 'Exploiting Timely Demand Information to Reduce Inventories', *European Journal of Operational Research*, 92(2) (1996): 239–53; F. Chen, Z. Drezner, J.K. Ryan and D. Simchi–Levi, 'Quantifying the Bullwhip Effect in a Simple Supply Chain: The Impact of Forecasting, Lead Times and Information', *Management Science*, 46(3) (2000): 436–43. Also see Cachon and Fisher (2000) above.

10 M. Zetlin, 'Blockbuster Could Have Bought Netflix for $50 Million, but the CEO Thought It Was a Joke', INC,

20 September 2019, https://www.inc.com/minda-zetlin/netflix-blockbuster-meeting-marc-randolph-reed-hastings-john-antioco.html.

11 M. Agarwal, 'India's Quick Commerce Race: Blinkit On Top After 2023; Can Rivals Catch Up?', Inc42, 26 December 2023, https://inc42.com/features/indias-quick-commerce-race-blinkit-on-top-after-2023-can-rivals-catch-up/.

12 See Colvin (2015).

13 D. Bonnet and G. Westerman, 'The Best Digital Business Models Put Evolution before Revolution', *Harvard Business Review*, 20 (2015), http://search.ebscohost.com/login.aspx?direct=true%7B&%7Ddb=bth%7B&%7DAN=118648197%7B&%7Dsite=ehost-live.

14 The overall DT value is the sum of values across time, and a net present value perspective is appropriate to examine the value created over time.

15 M. Bensaou, 'Interorganizational Cooperation: The Role of Information Technology and Networks in an Empirical Comparison of U.S. and Japanese Supplier Relations', *Information Systems Research*, *8*(2) (1997), 107–24, https://doi.org/10.1287/isre.8.2.107.
16 O. Hart, and J. Moore, (1988), 'Incomplete Contracts and Renegotiation', *Econometrica: Journal of the Econometric Society*, 755–85.
17 O. Hart, and J. Moore, (1990), 'Property Rights and the Nature of the Firm', *Journal of Political Economy*, *98*(6), 1119–58.

Chapter 10: The Extended DaWoGoMo© Model

1 S. Brown, 'Google's Sundar Pichai Says Tech Is a Powerful Agent for Change', MIT Management Sloan School, 17 March 2022, https://mitsloan.mit.edu/ideas-made-to-matter/googles-sundar-pichai-says-tech-a-powerful-agent-change.
2 J. Engel, 'Why Does Culture "Eat Strategy for Breakfast"?, *Forbes*, 10 December 2021, https://www.forbes.com/sites/forbescoachescouncil/2018/11/20/why-does-culture-eat-strategy-for-breakfast/?sh=37ddf1a61e09.
3 See Sia et al. (2015), https://hbsp.harvard.edu/product/NTU071-PDF-ENG.
4 E.H. Schein, *Organizational Culture and Leadership* Vol. 2 (Hoboken, John Wiley & Sons, 2010).

Chapter 11: The Existential Purpose: Life and Digitization

1 'Behold the Most Complicated Object in the Known Universe', NPR, 26 February 2014, https://www.wnyc.org/story/michio-kaku-explores-human-brain/.

2 'Most Views for an Animal on YouTube', Guinness World Records, 13 March 2018, https://www.guinnessworldrecords.com/world-records/444693-most-views-for-an-animal-on-youtube.

3 'Most Popular YouTube Animal Videos of the Decade', House Fur, 3 January 2020, https://housefur.com/10-most-popular-animal-videos-of-the-decade/.

4 R. Dawkins, *The Selfish Gene* (Oxford, Oxford University Press, 2016).

5 S. Ghoshal and P. Moran, 'Bad for Practice: A Critique of the Transaction Cost Theory', *Academy of Management Review*, 21(1) (1996): 13–47.

6 A. Mas–Colell, M.D. Whinston and J.R Green, *Microeconomic Theory* Vol. 1 (New York: Oxford University Press, 1995).

7 Herbert A. Simon, 'A Behavioral Model of Rational Choice', *The Quarterly Journal of Economics*, 69(1) (1955): 99–118, https://doi.org/10.2307/1884852.

8 Note that not all philosophies may be religious. Some philosophies may have a complex intertwined relationship with different religions.

9 C. Darwin, *On the Origin of Species by Means of Natural Selection* (London, Murray, 1859).

10 C. Camerer, G. Loewenstein and D. Prelec, 'Neuroeconomics: How Neuroscience Can Inform Economics', *Journal of Economic Literature*, 43(1) (2005): 9–64. https://doi.org/10.1257/0022051053737843.

11 P.W. Glimcher, A. Rustichini, 'Neuroeconomics: the Consilience of Brain and Decision', *Science*, 306(5695) (2004): 447–452. https://doi.org/10.1126/science.1102566.

12 A.R. Caggiula and B.G Hoebel, '"Copulation-Reward Site" in the Posterior Hypothalamus', *Science,* 153(3741) (1966): 1284–85; J. Mendelson, 'Lateral Hypothalamic Stimulation

in Satiated Rats: The Rewarding Effects of Self-Induced Drinking', *Science*, 157(3792) (1967): 1077–1079. Also see Camerer et al. (2005) above.

13 See Camerer et al. (2005) above.

14 R.C. O'Reilly and Y. Munakata, *Computational Explorations in Cognitive Neuroscience: Understanding the Mind by Simulating the Brain* (Cambridge, MA: MIT Press, 2000); E.O. Voit, *Computational Analysis of Biochemical Systems: A Practical Guide for Biochemists and Molecular Biologists* (Cambridge, UK: Cambridge University Press, 2000).

15 For example, matter is reduced into fundamental particles, comprising electrons, protons and their interrelationships. Also see O'Reilly and Munakata (2000) above.

16 F. Crick, 'Of Molecules and Men', *Nature* 219(5155) (1966): 758–62; E.E. Schadt and J.L.M. Björkegren, 'NEW: The Next Generation of Gene Expression Studies' *Trends in Genetics* 28(12) (2012): 527–535.

17 See Crick (1966) above, 11.

18 M. Gazzaniga, R. Ivry, R. George and G. Magnum, *Cognitive Neuroscience: The Biology of the Mind* (Cambridge, MA: MIT Press, 1998).

19 A. Dimoka, P.A. Pavlou and F.D. Davis, 'Research Commentary—NeuroIS: The Potential of Cognitive Neuroscience for Information Systems Research', *Information Systems Research*, 22(4) (2011): 687–702.

20 D.C. Van Essen et al., 'The Brain Analysis Library of Spatial Maps and Atlases (BALSA) Database', *Neuroimage*, 144 (2017): 270–274. https://doi.org/10.1016/j.neuroimage.2016.04.002.

21 T. Kalil, W. Koroshetz and G.K. Farber, 'NIH-Funded Scientists Identify 97 Previously Unknown Regions of the Brain', Obama White House Archives, 25 July 2016, https://obamawhitehouse.archives.gov/blog/2016/07/25/nih-funded-

scientists-identify-97-previously-unknown-regions-brain. Also see: M.F. Glasser et al., 'A Multi-Modal Parcellation of Human Cerebral Cortex', *Nature*, July 20 (2016). 171–78.
22 G. Morse, 'Decisions and Desire', *Harvard Business Review*, 84(1) (2006): 42.
23 See Camerer et al. (2005) above.
24 For example, Deppe et al. (2005) studied the brain regions and choices of different consumer goods brands. An fMRI-based examination of twenty-two participants choosing nearly indistinguishable brands found low activations of working memory and reasoning only when the target brand was the already-favourite brand, and brain regions assessing aspiration and self-reflection became more activated during decision-making. Also see Dimoka et al. (2011) above.
25 For example:
   a) Reward and utility are associated with the nucleus accumbens, caudate nucleus and putamen. Also see S.M. McClure, D.I. Laibson, G. Loewenstein and J.D. Cohen, 'Separate Neural Systems Value Immediate and Delayed Monetary Rewards', *Science*, 306(5695) (2004): 503–07. https://doi.org/10.1126/science.1100907.
   b) Prefrontal cortex and anterior cingulate cortex are associated with calculations. See M. Ernst and M.P. Paulus, 'Neurobiology of Decision Making: A Selective Review from a Neurocognitive and Clinical Perspective', *Biological Psychiatry*, 58(8) (2005): 597–604.
   c) Fear is associated with the amygdala. See J. LeDoux, 'The Emotional Brain, Fear and the Amygdala', *Cellular and Molecular Neurobiology*, 23(4-5) (2003): 727–38.
   d) Moral judgement is associated with the posterior superior temporal sulcus and frontopolar cortex. See J.S. Borg et al. 'Consequences, Action and Intention as Factors in Moral

Judgements: An fMRI Investigation', *Journal of Cognitive Neuroscience*, 18(5) (2006): 803–17.
e) See Dimoka et al. (2011).
f) It is important to note that there are many-to-many mappings between mental processes and brain regions. That is, a mental process may activate different brain regions and vice versa. Also see Dimoka et al. (2011).

26 Also see S. Edelman, *Computing the Mind: How the Mind Really Works* (Oxford, NY: Oxford University Press, 2008).

27 L.W. Barsalou, 'Perceptions of Perceptual Symbols', *Behavioral and Brain Sciences* 22(4) (1999): 637–60. https://doi.org/ https://doi.org/10.1017/S0140525X99532147.

28 Also see Edelman (2008).

29 Ibid.

30 See R.P. Abelson, 'Psychological Status of the Script Concept', *American Psychologist*, 36(7) (1981): 715–29. https://doi.org/ http://dx.doi.org/10.1037/0003-066X.36.7.715; D.A. Gioia, 'Symbols, Scripts and Sensemaking: Creating Meaning in the Organizational Experience', in *The Thinking Organization*, eds. H.P. Sims Jr. and D.A. Gioia (San Francisco: Jossey–Bass, 1986), 49–74.

31 A.N. Mishra and R. Agarwal, 'Technological Frames, Organizational Capabilities and IT Use: An Empirical Investigation of Electronic Procurement', *Information Systems Research* 21(2) (2010): 249–70. https://doi.org/https://doi.org/10.1287/isre.1080.0220.

32 D. Rumelhart, 'Schemata: The Building Blocks of Cognition', in *Theoretical Issues in Reading Comprehension*, eds. R. Spiro, B. Bruce and W. Brewer (Hillsdale, NJ: Erlbaum), 33–58.

33 K.D. Elsbach, P.S. Barr and A.B. Hargadon, 'Identifying Situated Cognition in Organizations', *Organization Science* 16(4) (2005): 422–33, https://doi.org/10.1287/orsc.1050.0138;

F.B. Tan and M.G. Hunter, 'The Repertory Grid Technique: A Method for the Study of Cognition in Information Systems', *MIS Quarterly*, 26(1) (2002): 39–57; J.P. Walsh, 'Managerial and Organizational Cognition: Notes from a Trip Down Memory Lane', *Organization Science* 6(3) (1995): 280–321.

34 R. Kanwar, J.C. Olson and L.S. Sims, 'Toward Conceptualizing and Measuring Cognitive Structures', in *NA—Advances in Consumer Research*, eds. Kent B. Monroe (Vol. 8) (Ann Arbor, MI: Association for Consumer Research, 1981), 122–27.

35 See Edelman (2008).

36 Note: Throughout the rest of the book, the terms 'cognitive schema', 'individual' and 'life' are used interchangeably.

37 The implications of this are huge, and I leave it for you to imagine.

Chapter 12: Revisiting the Instrumental Purpose: Digital Transformation and Individuals

1 Kenneth Miller, 'How Our Ancient Brains Are Coping in the Age of Digital Distraction', Discover, 22 April 2020, https://www.discovermagazine.com/mind/how-our-ancient-brains-are-coping-in-the-age-of-digital-distraction.

2 J. Anderson and L. Rainie, 'Improvements Ahead: How Humans and AI Might Evolve Together in the Next Decade', Pew Research Center, 10 December 2018, https://www.pewresearch.org/internet/2018/12/10/improvements-ahead-how-humans-and-ai-might-evolve-together-in-the-next-decade/.

3 A.C. Kamil and K. Cheng, 'Way-Finding and Landmarks: The Multiple-Bearings Hypothesis', *Journal of Experimental Biology*, 204(1) (2001): 103–13.

4 J.E. Rouse, *Respiration and Emotion in Pigeons* (Montana, Kessinger Publishing, 1905).

5 We will discuss more about the unique human existential purpose in Chapters 14 and 15; see Edelman (2008); see LeDoux (2003); M.S. Gazzaniga, 'Right Hemisphere Language Following Brain Bisection: A 20-Year Perspective', *American Psychologist* 38(5) (1983): 525–37. https://doi.org/10.1037/0003-066X.38.5.525.
6 V.R. Lakshminarayanan and L.R. Santos, 'Capuchin Monkeys Are Sensitive to Others' Welfare', *Current Biology*, 18(21) (2008): R999–R1000.
7 See LeDoux (1998), 11.
8 M. Lewis, 'The Emergence of Human Emotions', *Handbook of Emotions* 304 (2008).
9 However, life's emotional purpose is complex. We don't want to be afraid or scared but may want to stimulate fear by watching a horror movie. We may want to be satiated by eating food when hungry but may want to be food-deprived when dieting.
10 W.J. Havlena, M.B. Holbrook and D.R. Lehmann, 'Assessing the Validity of Emotional Typologies', *Psychology & Marketing*, 6(2) (1989): 97–112.
11 W.J. Havlena and M.B. Holbrook, 'The Varieties of Consumption Experience: Comparing Two Typologies of Emotion in Consumer Behaviour', *Journal of Consumer Research*, 13(3) (1986): 394–404.
12 Defining emotions has been so challenging that some have even suggested that emotions may not (or need not) be defined; see L. Pessoa, *The Cognitive-Emotional Brain: From Interactions to Integration* (Cambridge, MA: MIT Press, 2013); also see J. LeDoux, *The Emotional Brain: The Mysterious Underpinnings of Emotional Life* (Simon and Schuster, 1998).
13 Scherer, K.R., 'What Are Emotions? And How Can They Be Measured?', Social Science Information, 44, 2005, 693–727. https://doi.org/10.1177/0539018405058216.

14 Emotions are not a theoretical or nebulous concept. Well-developed methods make it easier to identify emotions. Various instruments used to assess emotions include the Self-Assessment Manikin (SAM). SAM relies on non-verbal pictorial assessment as it examines an individual's affective reaction—measuring the reaction's pleasure, arousal and dominance—in response to a stimulus. See J.A. Russell and A. Mehrabian, 'Evidence for a Three-Factor Theory of Emotions. *Journal of Research in Personality, 11*(3), (1977), 273–94. https://doi.org/10.1016/0092-6566(77)90037-X. Similarly, a Semantic Differential Scale examines the three dimensions—pleasure, arousal and dominance—through a rating of eighteen bipolar adjective pairs, such as unhappy–happy and relaxed–stimulated. Indeed, various measures are being used by marketers and organizations, such as to map consumption-related emotions. AI-based analytics and sentiment analysis models are enabling many of these endeavours; also see A. Bechara and A.R. Damasio, 'The Somatic Marker Hypothesis: A Neural Theory of Economic Decision', *Games and Economic Behavior*, 52(2) (2005): 336–72.

15 Just as were pigeons (see above). Also see J.J. Prinz, *Gut Reactions: A Perceptual Theory of Emotion.* (Oxford, Oxford University Press, 2004).

16 G. Small and G. Vorgan, 'Your iBrain: How Technology Changes the Way We Think', *Scientific American*, 1 October 2008, https://www.scientificamerican.com/article/your-ibrain/.

17 Note that previous literature often used affect and emotion interchangeably; also see M. Beauregard, J. Levesque and P. Bourgouin, 'Neural Correlates of Conscious Self-Regulation of Emotion', *The Journal of Neuroscience* 2001, *21*(18), RC165; see L. Pessoa, 'On the Relationship between Emotion and Cognition', *Nature Reviews Neuroscience*, 9(2) (2008): 148.

18 Split-brain experiments rely on the fact that the brain is organized into two hemispheres that are connected with the corpus callosum—the brain fibre system that allows communication between the two halves. Neuroscientists sever this part to treat patients suffering from severe epilepsy (see Gazzaniga, 1983). Gazzaniga shows how the left brain hemisphere—responsible for language—limits the ability of an individual to talk about the stimulus recognized by the right-hand brain, even though the right-hand brain can respond to the stimulus (without being able to speak about it) when the connection between the left and right-hand brains is severed (Gazzaniga, 1983). Further, brain-split experiments have shown that the right hemisphere evaluates emotions and can transfer these across to the left hemisphere, even when it is not able to send the details of the thinking, including the stimulus, to the left brain (see LeDoux, 1998); see Gazzaniga (1983), 525; also see Pessoa (2008) above.

19 A. Bechara and A.R. Damasio, 'The Somatic Marker Hypothesis: A Neural Theory of Economic Decision', *Games and Economic Behavior*, 52(2) (2005): 336–72; J.D. Greene et al., 'An fMRI Investigation of Emotional Engagement in Moral Judgement', *Science*, 293(5537) (2001): 2105–08; K.N. Ochsner, S.A Bunge, J.J. Gross and J.D. Gabrieli, 'Rethinking Feelings: An fMRI Study of the Cognitive Regulation of Emotion', *Journal of Cognitive Neuroscience*, 14(8) (2002): 1215–29; A.G. Sanfey et al. 'The Neural Basis of Economic Decision-Making in the Ultimatum Game', *Science*, 300(5626) (2003): 1755–58.

20 Although several parts of the limbic system have been empirically found to play a crucial role in emotional processes, there is considerable ambiguity about the general brain system associated with emotional processes. For example, in

documenting the historical perspective on the topic, Pessoa (2008) underlines the earlier works by Papez (J.W. Papez, A Proposed Mechanism of Emotion. Archives of Neurology & Psychiatry, 38, 725–743, (1937), https://doi.org/10.1001/archneurpsyc.1937.02260220069003) and Maclean (P.D. Maclean, 'Psychosomatic Disease and the "Visceral Brain"; Recent Developments Bearing on the Papez Theory of Emotion', Psychosomatic Medicine, 11, 338–53, (1949), https://doi.org/10.1097/00006842-194911000-00003). The former identified the anatomical foundations of emotions as a circuit involving the hypothalamus, hippocampal formation, cingulate cortex and their interconnections. These led to the concept of the limbic system, which identified various brain areas (such as the hippocampus). Pessoa argued that these are no longer considered important for emotional brain processes. Further, Pessoa argues that some other areas previously considered unimportant (such as the orbitofrontal cortex) are known to play a crucial role in emotional processing. Also, there is a lack of consensus about the limbic system's anatomy and functions, as it has been linked to diverse functions, such as motor responses, cognitive and sensory processes and learning (Pessoa, 2008). Despite the concerns about the unclear generic system for processing emotions, there is considerable evidence about the specific neural correlates for various emotions.

21 While the research in the domain is in its infancy, the researchers studying affect have unravelled complex neural circuitry regulating emotions, involving the orbital frontal cortex, amygdala, anterior cingulate cortex and several other interconnected regions (R.J. Davidson, K.M. Putnam and C.L. Larson, 'Dysfunction in the Neural Circuitry of Emotion Regulation—A Possible Prelude to Violence', *Science*, 289, 591–94, (2000), http://dx.doi.org/10.1126/science.289.5479.591).

However, there are no specific parts of the brain that trigger the arousal of emotions in particular and each emotion (say, fear and pleasure) may involve the use of different circuitry (see LeDoux 1998 above). LeDoux argues that emotions and cognition are not associated with specific functional areas in the brain and may activate many similar regions of the brain, as there is no dedicated area for computing emotions or cognition. Other neuroscientists echo the findings (Pessoa, 2008). In addition, the two regulate each other, with cognition influencing emotions and vice versa (see K.S. Blair, B.W. Smith, D.G.V. Mitchell, J. Morton, M. Vythilingam, L. Pessoa, D. Fridberg, A. Zametkin, D. Sturman, E.E. Nelson, W.C. Drevets, D.S. Pine, A. Martin and R.J.R Blair, 'Modulation of Emotion by Cognition and Cognition by Emotion. *NeuroImage*, 35(1), 430–40, (2007), https://doi.org/10.1016/j.neuroimage.2006.11.048.

22 Specifically, the trolley task assesses whether individuals were asked if they would choose to do something that would lead to the death of a different number of people, from those that may die if they don't do anything. See Greene et al. (2001) above.

23 S. Brown and X. Gao, 'The Neuroscience of Beauty', *Scientific American*, 27 (2011); S. Zeki, J.P. Romaya, D.M. Benincasa and M.F. Atiyah, 'The Experience of Mathematical Beauty and Its Neural Correlates', *Frontiers in Human Neuroscience*, 8,68, (2014).

24 A.J. Blood and R.J. Zatorre, 'Intensely Pleasurable Responses to Music Correlate with Activity in Brain Regions Implicated in Reward and Emotion', *Proceedings of the National Academy of Sciences*, 98(20) (2001): 11818–23.

25 In adults and infants, the experimental arousal of positive, approach-related emotions is associated with selective activation of the left frontal region, while the arousal of negative, withdrawal-related emotions is associated with

selective activation of the right frontal region. Individual differences in baseline measures of frontal asymmetry are associated with dispositional mood, affective reactivity, temperament and immune function. These studies suggest that neural systems mediating approach and withdrawal-related emotion and action are, in part, represented in the left and right frontal regions, respectively, and that individual differences in the activation levels of these systems are associated with a coherent nomological network of associations that constitute a person's affective style. (39) R.J. Davidson, (1992), 'Emotion and Affective Style: Hemispheric Substrates'.

26 See LeDoux (1998).
27 C. Darwin, *The Origin of Species* (New York, NY: Bantam Classics; Reissue edition, 1999).
28 D. Konstan, 'Aristotle on Anger and the Emotions: The Strategies of Status', *Ancient Anger: Perspectives from Homer to Galen* (2003): 99–120.
29 See Ochsner et al. (2002).
30 Specifically, neural correlates of reappraisal include lateral and medial prefrontal regions, amygdala and media orbito-frontal cortex. While the prefrontal regions realize positive activation, there is a decrease in action in the amygdala and media orbito-frontal cortex (see Ochsner et al., 2002). A detailed analysis of the neural circuitry involved in reappraisal is now available (see Ochsner et al. 2002).
31 Evolutionary biologists use fitness to model the relationship between genetic code (the antecedent) and fitness (the consequence) See Per Bak and Kim Sneppen, 'Punctuated Equilibrium and Criticality in a Simple Model of Evolution', *Physical Review Letters* 71 (1993): 4083. According to evolutionary biologists, greater fitness is triggered by genetic mutations. Theoretically, a single spin move in the genetic

code represents such a mutation. Building on the fitness logic, organizational and social science scholars use fitness landscapes to study the adaptation of organizations. See Daniel A. Levinthal, 'Adaptation on Rugged Landscapes', *Management Science* 43(7):934–50, (1997), https://doi.org/10.1287/mnsc.43.7.934

32 Organizations focus on choices for better performance as well. Using biologists' logic, organizational researchers examine how a change in organizational elements (such as strategies) influences performance (fitness). In fact, there is empirical evidence for these claims. The changes in irrigation works and agricultural strategies of Balinese farmers influence the fitness of their agrarian ecosystem; J.S. Lansing and J.N. Kremer, 'Emergent Properties of Balinese Water Temple Networks: Coadaptation on a Rugged Fitness Landscape', *American Anthropologist* 95, no. 1 (1993): 97–114; Bill McKelvey, (1999), 'Avoiding Complexity Catastrophe in Coevolutionary Pockets: Strategies for Rugged Landscapes', *Organization Science* 10(3): 294–321.

33 See 211.

34 Judgements help evaluate what is right or not and offer a very intelligent perspective for doing so. Robinson (2005) provides a nuanced distinction between emotions and their associated judgements. For example, there is a difference between trembling feelings caused by climbing stairs and trembling in love caused by hearing the voice of one's loved one. Further, similar outcomes may be associated with different emotions. For example, shame and embarrassment may both lead to withdrawal and hiding behaviours; J. Robinson, *Deeper Than Reason: Emotion and Its Role in Literature, Music and Art*. (Oxford, Oxford University Press, 2005).

35 C.S. Carver and M.F. Scheier, 'Principles of Self-Regulation: Action and Emotion', in E.T. Higgins and R.M. Sorrentino,

eds., *Handbook of Motivation and Cognition: Foundations of Social Behavior* (New York, The Guilford Press, 3–52, (1990).
36 J. J. Gross, 'The Emerging Field of Emotion Regulation: An Integrative Review', *Review of General Psychology*, *2*(3), 271–99(1998), https://doi.org/10.1037/1089-2680.2.3.271.
37 C.K. Hsee, R.P. Abelson and P. Salovey, (1991), 'The Relative Weighting of Position and Velocity in Satisfaction', *Psychological Science*, 2(4), 263–67; C.K. Hsee, R.P. Abelson and P. Salovey, (1991), 'Velocity Relation: Satisfaction as a Function of the First Derivative of Outcome over Time', *Journal of Personality and Social Psychology*, 60(3), 341.
38 J. Goudreau, 'IBM CEO Predicts Three Ways Technology Will Transform the Future of Business', *Forbes*, 8 March 2013, https://www.forbes.com/sites/jennagoudreau/2013/03/08/ibm-ceo-predicts-three-ways-technology-will-transform-the-future-of-business/?sh=915e6df6a732.
39 'Technology and Human Vulnerability', *Harvard Business Review*, September 2003, https://hbr.org/2003/09/technology-and-human-vulnerability.
40 S. Mitter, 'Facebook Becomes Youngest Company to Hit $1 Trillion Valuation', YOURSTORY, 30 June 2021, https://yourstory.com/2021/06/facebook-valuation-1-trillion-youngest-company-silicon-valley.
41 Numbers as of August 2023. See S.J. Dixon, "Facebook: Quarterly Number of MAU (Monthly Active Users) Worldwide 2008–2023', Statista, 21 May 2024, https://www.statista.com/statistics/264810/number-of-monthly-active-facebook-users-worldwide/.
42 Numbers as of April 2023. See A. Petrosyan, 'Worldwide Digital Population 2024', Statista, 22 May 2024, https://www.statista.com/statistics/617136/digital-population-worldwide/.

43 S.J. Dixon, 'Average Daily Time Spent on Social Media Worldwide 2012–2024', Statista, 10 April 2024, https://www.statista.com/statistics/433871/daily-social-media-usage-worldwide/.

44 S. Pichai, 'Investing in India's Digital Future', Google Blogs, 13 July 2020, https://blog.google/inside-google/company-announcements/investing-in-indias-digital-future/.

45 A. Petrosyan, 'Transactional e-Government Service Offerings 2014–2022', Statista, 1 December 2022, https://www.statista.com/statistics/421610/global-transactional-government-website-services/.

46 S. Brown, 'Google's Sundar Pichai Says Tech Is a Powerful Agent for Change', MIT Management Sloan School, 17 March 2022, https://mitsloan.mit.edu/ideas-made-to-matter/googles-sundar-pichai-says-tech-a-powerful-agent-change.

47 T. Gerken, 'Bill Gates: AI Is Most Important Tech Advance in Decades', BBC, 22 March 2023, https://www.bbc.com/news/technology-65032848.

Chapter 13: Operational Purpose Revisited: The Force Underlying DaWoGoMo© + Culture

1 Christopher Columbus, Royal Museums Greenwich, https://www.rmg.co.uk/stories/topics/christopher-columbus#:~:text=What%20did%20Columbus%20aim%20to,cargoes%20of%20silks%20and%20spices.

2 J.G. March, 'Exploration and Exploitation in Organizational Learning', *Organization Science*, 2(1) (1991): 71–87.

3 P. Setia, M. Setia, R. Krishnan and V. Sambamurthy, 'The Effects of the Assimilation and Use of It Applications on Financial Performance in Healthcare Organizations', *Journal of the Association for Information Systems*, 12(3) (2011): 1.

4 A. Wittenberg–Cox, 'Uber's Gender Pay Gap Study May Show the Opposite of What Researchers Were Trying to Prove', *Forbes*, 24 September 2018, https://www.forbes.com/sites/avivahwittenbergcox/2018/09/23/gender-paygap-uber-case-study/?sh=4d29bbc4b555.

Chapter 14: The Existential Purpose: Computational Freedom

1 R. Feloni, 'Billionaire LinkedIn Founder Reid Hoffman Says His Masters in Philosophy Has Helped Him More than an MBA', *Business Insider*, 24 November 2017, https://www.businessinsider.in/billionaire-linkedin-founder-reid-hoffman-says-his-masters-in-philosophy-has-helped-him-more-than-an-mba/articleshow/61786776.cms.
2 Ibid.
3 Ibid.
4 L.R. Goldberg, 'An Alternative "Description of Personality": The Big-Five Factor Structure', in *Personality and Personality Disorders* (London, Routledge, 2013), 34–47.
5 B.E. Stein, M.T. Wallace and T.R. Stanford, (1999). 'Development of Multisensory Integration: Transforming Sensory Input into Motor Output', *Mental Retardation and Developmental Disabilities Research Reviews*, 5(1), 72–85; R.J. Dolan, J.S. Morris and B. de Gelder, (2001), 'Crossmodal Binding of Fear in Voice and Face', *Proceedings of the National Academy of Sciences*, 98(17), 10006–110; J.L. Armony and R.J. Dolan, 'Modulation of Auditory Neural Responses by a Visual Context in Human Fear Conditioning', *Neuroreport* 12, no. 15 (2001): 3407–11, https://doi.org/10.1097/00001756-200110290-00051; P.J. Laurienti, A.S., Burdette, J.H., Maldjian, J.A., Y.-F. Yen and D.M. Moody, (2002),

'Dietary Caffeine Consumption Modulates fMRI Measures', *Neuroimage*, *17*(2), 751–57.

6   R.L. Peterson, 'The Neuroscience of Investing: fMRI of the Reward System', *Brain Research Bulletin*, 67(5) (2005): 391–97, https://doi.org/10.1016/j.brainresbull.2005.06.015.

7   In fact, reward dynamics are universal, going even beyond the human species and guiding the behaviours of most living species. Many living beings (including humans) have a central nervous system; however, the development of centralized areas of nerve cells differs across species. Indicating its prevalence in the universe, the pursuit of reward goes beyond the living beings with a nervous system. The unicell organisms also pursue rewards. Even the earliest aquatic cells move to search for resource-rich regions, as this helps them reproduce and prosper.

8   Dopamine (DA) neurons activated in monkeys when the animal was presented with food (W. Schultz and R. Romo, 'Dopamine Neurons of the Monkey Midbrain: Contingencies of Responses to Stimuli Eliciting Immediate Behavioral Reactions', *Journal of Neurophysiology, 63*(3), 607–24, (1990). Also, Morse, 2006 underlines, 'Produced in the ancient structures of our animal brains, it helps to regulate the brain's appetite for rewards and its sense of how well rewards meet expectations.' (46).

9   D.M. Small et al., 'Changes in Brain Activity Related to Eating Chocolate: From Pleasure to Aversion', *Brain*, 124(9) (2001): 1720–1733.

10 I. Aharon et al., 'Beautiful Faces Have Variable Reward Value: fMRI and Behavioural Evidence', *Neuron*, 32(3) (2001): 537–51.

11 M. Hsu et al. (M. Hsu, M. Bhatt, R. Adolphs, D. Tranel, and C.F. Camerer, 'Neural Systems Responding to Degrees of Uncertainty in Human Decision-Making', *Science*, 310, 1680–83(2005), http://dx.doi.org/10.1126/science.1115327)

underline that activation in the caudate nucleus is greater when the magnitude of the anticipated reward is high. Its level of activation for anticipated losses is also different from that of anticipated gains. See also M.R. Delgado, M.M. Miller, S. Inati, E.A. Phelps, 'An fMRI Study of Reward-Related Probability Learning', *Neuroimage*, 2005 Feb 1;24(3):862-73, doi: 10.1016/j.neuroimage.2004.10.002. Epub 2004 Nov 18. PMID: 15652321.

12 S.M. McClure, D.I. Laibson, G. Loewenstein and J.D. Cohen, 'Separate Neural Systems Value Immediate and Delayed Monetary Rewards', *Science*, 306(5695) (2004): 503–07, https://doi.org/10.1126/science.1100907; A.E. Kelley and K.C. Berridge, 'The Neuroscience of Natural Rewards: Relevance to Addictive Drugs', *Journal of Neuroscience*, 22(9) (2002): 3306–11.

13 R.L. Peterson (2005, 391) unravels five major dopamine pathways through which dopaminergic neurons carry information across brain regions.

14 T.S. Braver et al., 'A Parametric Study of Prefrontal Cortex Involvement in Human Working Memory', *Neuroimage*, 5(1) (1997): 49–62.

15 J.D. Cohen et al., 'Temporal Dynamics of Brain Activation during a Working Memory Task', *Nature*, 386(6625) (1997): 604–08.

16 Michael O'Shea, *The Brain: A Very Short Introduction* (Oxford: Oxford University Press, 2005), 136 pages, ebook.

17 Ibid.

18 H.A. Simon, (1957), *Models of Man* (New York, Wiley, 5).

19 James G. March, 'Bounded Rationality, Ambiguity, and the Engineering of Choice', *The Bell Journal of Economics* 9, no. 2 (1978): 587–608, https://doi.org/10.2307/3003600.

20 See Edelman (2008).

21 J. March (1978) above (4).

22 R. Kurzweil, (2013), *How to Create a Mind: The Secret of Thought Revealed* (Penguin Books).

23 J.R. Bettman, (1979), 'Memory Factors in Consumer Choice: A Review', *The Journal of Marketing*, 37–53.

24 A. Newell and H.A. Simon, (1972), *Human Problem Solving* (Englewood Cliffs: Prentice Hall), http://garfield.library.upenn.edu/classics1980/A1980KD04600001.pdf; H. Simon, (1969), 'The Sciences of the Artificial', (Cambridge, MA, MIT Press); Roger N. Shepard, 'Recognition Memory for Words, Sentences, and Pictures', *Journal of Verbal Learning and Verbal Behavior* 6 (1967): 156–63.

25 T. Shallice and P. Burgess, (1996), 'The Domain of Supervisory Processes and Temporal Organization of Behaviour', *Phil. Trans. R. Soc. Lond. B*, *351*(1346), 1405–12, https://doi.org/10.1098/rstb.1996.0124.

26 J.A. Bargh, M. Chen, and L. Burrows, 'Automaticity of Social Behavior: Direct Effects of Trait Construct and Stereotype Activation on Action', *Journal of Personality and Social Psychology*, 71, 230–44, (1996), http://dx.doi.org/10.1037/0022-3514.71.2.230; T.L. Chartrand and J.A. Bargh, 'The Chameleon Effect: The Perception–Behavior Link and Social Interaction', *Journal of Personality and Social Psychology*, 76(6), 893–910, (1999), https://doi.org/10.1037/0022-3514.76.6.893; R.M. Shiffrin and W. Schneider, (1977), 'Controlled and Automatic Human Information Processing: Ii. Perceptual Learning, Automatic Attending and a General Theory', *Psychological Review*, *84*(2), 127.

27 J. Ronson, 'How One Stupid Tweet Blew Up Justine Saco's Life', *New York Times*, 12 February 2015, https://www.nytimes.com/2015/02/15/magazine/how-one-stupid-tweet-ruined-justine-saccos-life.html.

28 C. Camerer, G. Loewenstein and D. Prelec, 'Neuroeconomics: How Neuroscience Can Inform Economics', *Journal of Economic Literature*, 43(1) (2005): 9–64.
29 R. Aljafari, F. Soh, P. Setia and R. Agarwal, 'The Local Environment Matters: Evidence from Digital Healthcare Services for Patient Engagement', *Journal of the Academy of Marketing Science*, (2023): 1–23.
30 'Why Self-Driving Cars Must Be Programmed to Kill', *Technology Review*, 22 October 2015, https://www.technologyreview.com/2015/10/22/165469/why-self-driving-cars-must-be-programmed-to-kill/.
31 In fact, there is a history of thinking about artificial intelligence, as outlined by the noted thinker and Nobel laureate, Herbert Simon, in his works. See H.A Simon, *The Sciences of the Artificial* (Cambridge, MA: MIT Press, 1996).
32 'Sheryl Sandberg's Commencement address', MIT News, 8 June 2018, https://news.mit.edu/2018/sheryl-sandberg-commencement-address-0608.
33 There is a reason why children are encouraged to go out and play and ponder, and child labour is a crime.
34 V.S Ramachandran, '3 Clues to Understanding Your Brain', TED, March 2007, https://www.ted.com/talks/vs_ramachandran_3_clues_to_understanding_your_brain/transcript.

Chapter 15: Digital Organization

1 G. Brassil, 'Here's What Facebook's Sheryl Sandberg Told New Grads at MIT', CNBC.com, 8 June 2018, https://www.cnbc.com/2018/06/08/facebook-coo-sheryl-sandberg-commencement-speech-full-transcript.html.

2 See Gates (2023), https://www.gatesnotes.com/The-Age-of-AI-Has-Begun.
3 TED, 'Google's Driverless Car', YouTube video, 31 March 2011, https://www.youtube.com/watch?v=bp9KBrH8H04&t=2s.
4 In fact, this may be a good time to think about the various intense computations you perform on a daily basis. Right now, as you go through this book, you are reading and interpreting my words and linking them with a lot of things in your mind, combining them with a lot of knowledge you have already acquired so far in life. You have a certain meaning for the words; your brain is also processing my writing style, the language, the tone—all intense computational activities.
5 Also, consider that an autonomous car is equipped with driving capability equivalent of hundreds of years of driving experience.
6 Jennifer Coopersmith, 'The Laws of Thermodynamics: Thomson and Clausius', Oxford Academic, May 2015, https://academic.oup.com/book/8337/chapter-abstract/153993478?redirectedFrom=fulltext.
7 The notion of isolated systems is complex. A refrigerator may preserve or reduce the disorder of contents inside by freezing them, but it gives out heat, uses power produced in electric plants that generate a lot of heat and so on. Overall, a closed system that includes everything results in a higher level of entropy. Alternatively, if the universe is considered a controlled system, its overall entropy always tends to increase. Further, the idea of a spontaneous process implies that a refrigerator may create greater order and reduces entropy, but this is not a spontaneous process as work has to be done to counter the entropic tendency for foods to decay or achieve a state of greater disorder. Overall, work is required to counter entropy.

8 Beyond these examples, entropy as a concept has been conceptualized and tested in other educational domains, such as statistical sciences and communication systems.
9 See Brassil (2018), https://www.cnbc.com/2018/06/08/facebook-coo-sheryl-sandberg-commencement-speech-full-transcript.html.
10 Unit economics entails calculating the unit (say, customer) level value of a certain offering—product or service—or business model.
11 Many retail organizations, such as Amazon, the upcoming ONDC or Ajio, have transformed the expectations of people and the organization of retail is a far cry from the days of the haat.
12 It is logical to think above revenues, sales, profits, net margins, EBITDA and so on. While organizations survive or exist to create value, organizational value is *only* a metric used to assess whether the organization is doing something meaningful that fulfils the individual's purpose—the realization of computational freedom. It is imperative to note that money is a measure of value. This aspect has been debated and underlined for a long time, going as far back as Aristotle. Aristotle pondered whether money is the entity that exchanges value. However, soon he rightly realized that money is a measure of value and not value itself (see S. Fleetwood, (1997), 'Aristotle in the 21st Century', *Cambridge Journal of Economics*, *21*(6), 729–44.

Chapter 16: The Purpose

1 'Technology and Human Vulnerability', *Harvard Business Review*, September 2003, https://hbr.org/2003/09/technology-and-human-vulnerability.

2 Consider your icons in sports, such as cricket, basketball, athletics, football and tennis; in culture and the arts, such as singers, actors and dancers; in religions, such as Hinduism, Sikhism, Islam and Christianity; in society, politics, public agencies, writing, academics and business, such as founders, technology icons, professionals, etc.
3 See Dimoka et al. (2011).
4 'Entering the Trust Age', BlaBlaCar, https://blog.blablacar.com/trust
5 When the records of erstwhile plantations were unveiled, the names of humans and their *value* surfaced. The visceral burden of slavery was reduced in the nineteenth century (through initiatives such as the freedom of colonies and the slavery abolition act of 1833 that freed 8,00,000 Africans treated as British slave owners' property). However, the loss of freedom for many persists even today, though perhaps not that openly. Estimates indicate that slavery influenced 40.3 million people (per the UN data) in 2016, as 24.9 million were in forced labour and 15.4 million were married out of their wishes. Even today, the statistics overlay examples of some of the most immense, stark and visceral practices restricting freedom across sectors of the economy, such as domestic work, construction and agriculture. For example, the Human Rights Commission in Pakistan has revealed that there is extremely high mortality among children working in brick kilns, with approximately 5 per cent of families living close to these kilns having children who have lost their eyesight (See 'Bonded Labour', Al Jazeera, https://interactive.aljazeera.com/aje/2019/pakistan-bonded-labour/index.html). No nation is spared, as instances of slavery even in the 'developed' nations surface regularly.

'The History of British Slave Ownership Has Been Buried: Now Its Scale Can Be Revealed', *Guardian*, https://www.

theguardian.com/world/2015/jul/12/british-history-slavery-buried-scale-revealed. Also see https://www.ilo.org/sites/default/files/wcmsp5/groups/public/@dgreports/@dcomm/documents/publication/wcms_575479.pdf.

6 '10 Timeless Poems by Rabindranath Tagore', 6 May 2021, https://timesofindia.indiatimes.com/life-style/books/features/10-timeless-poems-by-rabindranath-tagore/photostory/75593222.cms?picid=75593254.

7 Philosophers and scholars from Aristotle on have talked about the geocentric model of the universe. However, the power of this *truth* changed how humans think about themselves and helped develop space capabilities—various rockets, space and satellite programmes. In India, the ancient texts of the Vedas underline early discoveries of the universe, planets and their movements around the sun and celestial phenomena such as eclipses being predicted through the sun temples.

8 See (O'Shea 2005) above.

9 Nicolelis Lab, http://www.nicolelislab.net/.

10 L. Hochberg, M. Serruya, G. Friehs et al., 'Neuronal Ensemble Control of Prosthetic Devices by a Human with Tetraplegia', *Nature* 442, 164–71 (2006). 'Monkey Controls Robotic Arm with Brain-Computer Interface' retrieved 29 January 2018, from http://scienceblogs.com/neurophilosophy/2008/05/29/monkey-controls-robotic-arm-wi/.

11 'Plant-based Foods Market to Hit $162 Billion in Next Decade, Projects Bloomberg Intelligence', *Bloomberg*, 11 August 2021, https://www.bloomberg.com/company/press/plant-based-foods-market-to-hit-162-billion-in-next-decade-projects-bloomberg-intelligence/.

12 J. Temple, 'Bill Gates: Rich Nations Should Shift Entirely to Synthetic Beef', *MIT Technology Review*, 14 February, 2021, https://www.technologyreview.com/2021/02/14/1018296/bill-

gates-climate-change-beef-trees-microsoft/; C. Elton, 'Plant-Based Meat Is the "Future", Billionaire Bill Gates Claims. What Has to Change?', euronews, 9 January 2023, https://www.euronews.com/green/2023/09/01/plant-based-meat-is-the-future-billionaire-bill-gates-claims-what-has-to-change.

13 'About Google', Google, https://about.google/.
14 The only non-US system that is accessed by over 1 billion users, Aadhaar is a purposeful digital transformation of the public sector that has leveraged the platform logic (through its India Stack) to reach the masses. Aadhaar provides a biometrically authenticated twelve-digit number that forms the basis for authentication and authorization for various services created through the API provided by India Stack.
15 The inclusion provided by the programme is exemplary, as beyond allowing the account holder to maintain a zero balance, the account also offers insurance coverage of Rs 1,00,000 (about $1500) and the ability to maintain an overdraft of Rs 5000 ($80) giving them access to small loan functionality.
16 A. Gupta and P. Auerswald, 'How India Is Moving Toward a Digital-First Economy', *Harvard Business Review*, 8 (2017), https://hbsp.harvard.edu/product/H040HB-PDF-ENG.

Scan QR code to access the
Penguin Random House India website